SpringerBriefs in Computer Science

SpringerBriefs present concise summaries of cutting-edge research and practical applications across a wide spectrum of fields. Featuring compact volumes of 50 to 125 pages, the series covers a range of content from professional to academic.

Typical topics might include:

- A timely report of state-of-the art analytical techniques
- A bridge between new research results, as published in journal articles, and a contextual literature review
- A snapshot of a hot or emerging topic
- An in-depth case study or clinical example
- A presentation of core concepts that students must understand in order to make independent contributions

Briefs allow authors to present their ideas and readers to absorb them with minimal time investment. Briefs will be published as part of Springer's eBook collection, with millions of users worldwide. In addition, Briefs will be available for individual print and electronic purchase. Briefs are characterized by fast, global electronic dissemination, standard publishing contracts, easy-to-use manuscript preparation and formatting guidelines, and expedited production schedules. We aim for publication 8–12 weeks after acceptance. Both solicited and unsolicited manuscripts are considered for publication in this series.

**Indexing: This series is indexed in Scopus, Ei-Compendex, and zbMATH **

Liangrui Peng • Ruijie Yan

Multilingual Text Recognition

A Deep Learning Approach

Liangrui Peng
Department of Electronic Engineering
Tsinghua University
Beijing, China

Ruijie Yan
Department of Electronic Engineering
Tsinghua University
Beijing, China

ISSN 2191-5768   ISSN 2191-5776 (electronic)
SpringerBriefs in Computer Science
ISBN 978-981-96-7897-6   ISBN 978-981-96-7898-3 (eBook)
https://doi.org/10.1007/978-981-96-7898-3

© The Editor(s) (if applicable) and The Author(s), under exclusive license to Springer Nature Singapore Pte Ltd. 2025

This work is subject to copyright. All rights are solely and exclusively licensed by the Publisher, whether the whole or part of the material is concerned, specifically the rights of translation, reprinting, reuse of illustrations, recitation, broadcasting, reproduction on microfilms or in any other physical way, and transmission or information storage and retrieval, electronic adaptation, computer software, or by similar or dissimilar methodology now known or hereafter developed.
The use of general descriptive names, registered names, trademarks, service marks, etc. in this publication does not imply, even in the absence of a specific statement, that such names are exempt from the relevant protective laws and regulations and therefore free for general use.
The publisher, the authors and the editors are safe to assume that the advice and information in this book are believed to be true and accurate at the date of publication. Neither the publisher nor the authors or the editors give a warranty, expressed or implied, with respect to the material contained herein or for any errors or omissions that may have been made. The publisher remains neutral with regard to jurisdictional claims in published maps and institutional affiliations.

This Springer imprint is published by the registered company Springer Nature Singapore Pte Ltd.
The registered company address is: 152 Beach Road, #21-01/04 Gateway East, Singapore 189721, Singapore

If disposing of this product, please recycle the paper.

Foreword

Optical character recognition (OCR) is a representative artificial intelligence technology. The Research Center for Intelligent Document and Image Information Processing at Tsinghua University has been devoted to the field of OCR research for four decades. Guided by information entropy theory for pattern recognition and machine learning, statistical approaches using high-dimensional discriminant feature representations were explored, which increased the cross entropy of features and patterns and finally provided solutions for the problem of large-scale character set Chinese character recognition. The related research includes offline and online Chinese handwriting recognition, multilingual printed text recognition, scene text recognition, and form recognition. For multilingual printed text recognition, the developed TH-OCR software products support Chinese, Japanese, Korean, English, Mongolian, Tibetan, Uighur/Kazakh/Kirghiz, and Arabic scripts and have been widely used in various applications.

This book entitled *Multilingual Text Recognition: A Deep Learning Approach* presents the authors' research work on the problem of multilingual text recognition in an open environment. The contents of this book can be regarded as a new extension of TH-OCR technologies with the recent emergence of deep learning. The entire book not only contains the basic knowledge and theoretical analysis of deep neural network-based machine learning methods but also provides detailed practical solutions for multilingual text recognition in images and videos. Readers can learn how to obtain deep representations and convert them into semantic representations to solve the difficult problem of multilingual text recognition. The content of this book will help college students, postgraduates, and researchers understand deep neural networks for pattern recognition, and help readers study recent progress in multilingual text recognition. It is a good reference book for students and researchers in the OCR research field.

Beijing, China
March 2025

Xiaoqing Ding

Preface

Languages and their writing systems are not only symbols of human civilization but also important carriers of social information. Linguists estimate that there are more than 6000 languages and more than 200 writing systems around the world. This diversity creates obstacles to cross-cultural communication. The story of the Tower of Babel in the Bible describes the fantasy of human beings without language barriers. In the era of mobile internet, the language barrier can be overcome by using cutting-edge technologies. People can capture multilingual text images, convert them into text codes, and use machine translation to obtain their meanings. Multilingual text recognition is an important technology in the optical character recognition (OCR) research field for converting multilingual images into text codes.

With the advent of deep learning, designing deep neural networks for multilingual text recognition is an interesting research topic. This book summarizes our research efforts over the last several years. The essential problem is to find efficient representations for sequence modeling of multilingual text images. This book mainly introduces primitive representation learning (PREN), which is different from previous methods, including CTC-decoding methods and encoder-decoder-based methods. This book also presents a multielement attention network (MEAN), which is an encoder-decoder-based method that uses an improved self-attention mechanism. Pren2D is further proposed by combining PREN and MEAN to achieve better performance for irregular multilingual scene text images. For long text line images, a variant of the CTC decoding method that uses both a dynamic temporal residual network (DTRN) and an attention rectification network (ARN) is also proposed. All these methods are integrated into the TH-DL system framework for multilingual text recognition, which has achieved leading performance on the leaderboard of the RRC-MLT-2019 competition task, etc.

We would like to thank all the institutions and individuals who developed the databases used in the research of this book, including handwriting datasets (IFN/ENIT Database of Handwritten Arabic Words, IAM English Handwriting Database, and RIMES French Handwriting Database), multilingual scene text datasets (RRC MLT 2019/2017), RCTW, and commonly used English scene text. Special thanks are due to Professor Volker Märgner (Institute for Communications

Technology, Technical University Braunschweig, Germany) and Dr. Nibal Nayef (MyScript, Nantes, France) for their support.

We are enormously grateful to Lanlan Chang, Jingying Chen, and Kamesh Senthilkumar of Springer Nature for their kind support and patience throughout the project.

The first author would like to thank her colleagues and former and current graduate students in the Department of Electronic Engineering at Tsinghua University for their collaboration and help. Special thanks are due to Professor Xiaoqing Ding, Professor Shengjing Wang, Professor Changsong Liu, and Professor Yali Li for their constant support and insightful advice. Special thanks are due to Shanyu Xiao, Gang Yao, Pei Tang, Ning Ding, and Kemeng Zhao for their contributions to this research.

Moreover, we would like to acknowledge the support of the National Natural Science Foundation.

Finally, we would like to thank both families for their support.

Beijing, China Liangrui Peng
March 2025 Ruijie Yan

Declarations

Competing Interests The authors have no competing interests to declare that are relevant to the content of this manuscript.

Contents

1	**Introduction**	1
	1.1 Multilingual Text Recognition	1
	1.2 Multilingual Scripts and Related Encoding Standards	2
	1.3 Multilingual Text Recognition Methods	3
	1.3.1 Traditional Methods	4
	1.3.2 Deep Learning-Based Methods	5
	1.4 Datasets	10
	1.4.1 Scene Text Datasets	10
	1.4.2 Handwriting Datasets	13
	1.5 Evaluation Metrics	14
	1.5.1 Evaluation Metrics for Text Recognition	14
	1.5.2 Evaluation Metrics for End-to-End Text Detection and Recognition	15
	1.6 Book Overview	16
	1.6.1 Problem Analysis	16
	1.6.2 Research Content	17
	1.6.3 Book Structure	19
	References	20
2	**Primitive Representation Learning**	27
	2.1 Introduction of Primitive Representation Learning	27
	2.2 Primitive Representations	30
	2.2.1 General Form of Primitive Representations	30
	2.2.2 Pooling Aggregator	31
	2.2.3 Weighted Aggregator	32
	2.3 Visual Text Representations	33
	2.4 Primitive Representation Learning Network	34
	2.4.1 Feature Extraction	34
	2.4.2 Primitive Representation Learning	35
	2.4.3 Training and Inference	36

2.5	Semantic-Guided Decoding		38
	2.5.1	Language Model with Attention Mask	39
	2.5.2	Fusion of Visual and Semantic Information	41
2.6	Discussion		42
	2.6.1	English Scene Text Recognition	42
	2.6.2	Multioriented Chinese Scene Text Recognition	49
	2.6.3	Arabic Handwriting Recognition	51
	2.6.4	Semantic-Guided Decoding Experiment	52
2.7	Summary		56
References			56

3 Multielement Attention Mechanism — 59

3.1	Introduction		59
3.2	Multielement Attention Mechanism		61
3.3	Multielement Attention Network		64
	3.3.1	Feature Extraction	65
	3.3.2	Orientational Positional Encoding	65
	3.3.3	Encoder and Decoder	66
3.4	Primitive Representation Learning with Multielement Attention		67
3.5	Discussion		69
	3.5.1	Experimental Settings	69
	3.5.2	Comparison of Different Types of Multielement Attention Mechanisms	69
	3.5.3	Effect of Orientational Positional Encoding	70
	3.5.4	Comparison with Existing Methods	71
	3.5.5	Visualization	71
3.6	Summary		74
References			75

4 Dynamic Temporal Residual Learning and Attention Rectification — 77

4.1	Introduction		77
4.2	Temporal Residual Learning		81
4.3	Dynamic Weights for Temporal Residual Connections		83
4.4	Attention Rectification		85
	4.4.1	Positional Encoding	86
	4.4.2	Character Spatial Constraints	87
	4.4.3	Attention Rectification Process	87
	4.4.4	Training and Inference	88
4.5	Discussion		89
	4.5.1	Dynamic Temporal Residual Learning	89
	4.5.2	Attention Rectification	97
4.6	Summary		100
References			102

5 TH-DL Multilingual Text Recognition System Framework 105
- 5.1 System Framework 105
- 5.2 Model Selection for Different Tasks 106
- 5.3 Multilingual Scene Text Recognition System 108
- 5.4 Arabic Video Subtitle Recognition System 112
- 5.5 Conclusion and Future Work 114
- References 114

Chapter 1
Introduction

Abstract Multilingual text recognition is an important research topic in artificial intelligence and computer vision. It provides a powerful tool for extracting text information from images in various applications, such as content-based video and image retrieval. This chapter addresses the challenging task of multilingual text recognition for images captured in an open environment. After a brief review of traditional methods and deep learning-based methods for multilingual text recognition, the content of this book is introduced including the proposed primitive representation learning method, among others.

Keywords Multilingual text recognition · Encoding standard · Deep learning · Connectionist temporal classification · Encoder-decoder · Primitive representation learning

1.1 Multilingual Text Recognition

In the era of globalization, multilingual texts play an important role as carriers of cultural, economic and social information dissemination. However, the diversity of multilingual scripts creates obstacles for cross-cultural communication. With technological advances in the research fields of computer vision (CV) and natural language processing (NLP), bridging the gap among different scripts has become possible. For multilingual text images, optical character recognition (OCR) is a useful technology that can extract textual information from images and output text codes that can be automatically analyzed and translated. Traditional OCR technologies are designed to process document images digitized by scanners. In recent years, there have been increasing demands to process images captured by cameras, mobile phones or other hand-held devices, especially in emerging application scenarios such as autonomous driving and robot navigation. Compared with images digitized by scanners, multilingual text images acquired in these unconstrained scenarios present more challenges to the OCR research field.

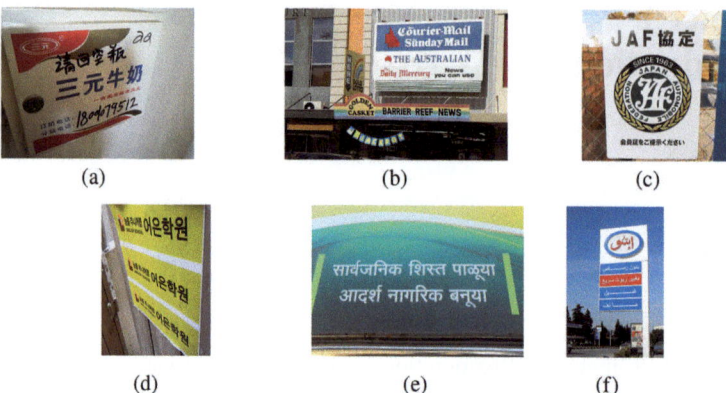

Fig. 1.1 Examples of multilingual text in natural scene images. (**a**) Chinese. (**b**) English. (**c**) Japanese. (**d**) Korean. (**e**) Hindi. (**f**) Arabic (From [81] with permission)

Some examples of multilingual text images captured in an open environment are shown in Fig. 1.1, demonstrating the main challenges in multilingual text recognition research.

1. **Diversity of text instances**. Text instances come in a variety of scripts, fonts or writing styles, sizes, colors, and layouts. In particular, the diversity of scripts results in different textual appearances and characteristics.
2. **Background diversity**. Backgrounds are complex in uncontrolled conditions. Some patterns in the background are similar to those in text; for example, the texture of a fence or railing looks like that of text instances.
3. **Diversity of imaging conditions**. Camera-captured text images often suffer from problems such as low resolution, uneven illumination, motion blur, distortion, occlusion and noise.

1.2 Multilingual Scripts and Related Encoding Standards

The task of multilingual text recognition is to convert the content in text images into text codes. It is necessary to analyze the characteristics of multilingual scripts and related encoding standards.

There are several major writing systems in the world, which can be categorized into alphabetic and nonalphabetic scripts. Alphabetic scripts use letters to represent individual sounds of languages. Although most modern alphabets, including the Latin alphabet, Cyrillic alphabet, and Arabic alphabet, are derived from the Phoenician alphabet invented 4000 years ago, different alphabets have their own characteristics. For example, the Arabic alphabet consists of 28 letters, most of which take three different shapes depending on the initial, medial, and final positions of a word. The Arabic script is semicursive with rich ligatures. An Arabic text line is written from right to left, and English words mixed in the Arabic text line are written from left to right. In contrast, Chinese script is representative of nonalphabet scripts. Chinese scripts consist of thousands of commonly used characters. Each Chinese character is a two-dimensional complex figure composed of strokes, which is a unity of shape, sound and meaning. These characteristics require more attention when corresponding text recognition methods are designed.

To facilitate the use of computers to store, transmit and share multilingual text information, developing text encoding standards for both alphabetic and nonalphabetic scripts is fundamental and important. An alphabet usually has a limited set of characters and symbols, which can be represented by a one-byte encoding scheme. Nevertheless, the encoding of commonly used Chinese characters requires two-byte forms, whereas the encoding of historical Chinese characters or rare characters in names and addresses requires four-byte forms.

A series of encoding standards have been developed over the past few decades. In 1963, the American Standards Association (ASA) released the first edition of the American Standard Code for Information Interchange (ASCII), which encodes the English alphabet in a 7-bit binary code format. The 7 bits of an ASCII character are stored as 8 bits in computers, with the leading bit being set to "0". For East Asian scripts with large character sets, native or local double-byte character set (DBCS) or multibyte character set (MBCS) encoding standards have been developed, including GB/T 2312-1980 for simplified Chinese, BIG 5 for traditional Chinese, SJIS based on JIS-X0208-1997 for Japanese and KSC 5601-1987 for Korean. These native or local CJK (Chinese, Japanese and Korean) character encoding standards adopt multibyte schemes by using both one-byte ASCII codes and double-byte character codes. The non-ASCII characters are represented by two successive bytes, with leading bits being set to "1". One inherent problem is that the double-byte character codes from different CJK encoding standards conflict.

The Unicode standard is a solution for mapping multiple character sets of the CJK languages into a single set of unified characters. In 1991, the International Organization for Standardization (ISO) Working Group and the newly established Unicode Consortium decided to work cooperatively to develop ISO/IEC 10646 and the Unicode standard in a compatible and synchronous manner. The first version of the Unicode standard, which includes 20,902 Chinese/Japanese/Korean idiographs, was released in October 1991. In the current version (15.0), Unicode defines 149,186 characters covering 161 modern and historic scripts. The widely used UTF-8 is an ASCII-compatible variable-length transformation format for Unicode, which is usually adopted for the output text of multilingual text recognition.

1.3 Multilingual Text Recognition Methods

Multilingual text recognition is related to pattern recognition, machine learning, computer vision and natural language processing. For printed document scanned images, traditional optical character recognition (OCR) commercial software products such as ABBYY FineReader, Kofax OmniPage and TH-OCR [84] have provided support for the recognition of multilingual document images. Printed documents usually have clean backgrounds and regular text regions, and the recognition process generally includes image preprocessing, layout analysis, text line segmentation, character/word segmentation and recognition, and postprocessing with a language model. However, these OCR systems usually do not support the recognition of scene text images captured in an open environment.

The recognition of multilingual text in the wild poses greater challenges, as there is more uncertainty in regard to the fundamental tasks of locating and identifying text. For human visual cognition, the processes of text localization and identification are usually interwoven. For computer vision algorithms, there are usually two independent or jointly optimized steps: text detection and text recognition. Text detection aims to output the bounding boxes of text regions in an input image, and text recognition predicts text codes in the detected regions.

This book focuses on the key problem of text recognition for multilingual scripts. The existing methods for multilingual text recognition can be divided into two categories: traditional methods and deep learning-based methods. Traditional multilingual text recognition methods are mainly based on handcrafted features, which are relatively sensitive to image quality and noise. With the advent of deep learning, various deep neural network-based methods have been proposed for multilingual text recognition [30, 52, 60, 69, 122]. Text recognition methods based on deep learning usually adopt sequence modeling schemes. Sequence-modeling methods can be divided into two main categories according to their decoding techniques: connectionist temporal classification (CTC) [36]-based methods and encoder-decoder-based methods. For encoder-decoder-based methods, RNN encoder-decoder or Transformer [105] encoder-decoder architectures are usually employed.

Recently, large vision language models (LVLMs), such as GPT-4o [1, 48], Gemini-1.5-Pro [104], Qwen2-VL [114], and InternVL2 [18], have played pivotal roles in advancing OCR technologies by incorporating contextual analysis into the recognition process [126]. However, it is not convenient to utilize LVLMs for large-scale OCR tasks in real applications because of computational burdens. Some recent OCR professional models including KOSMOS2.5 [72], TextMonkey [66], Florence [120], and GOT [115]. have provided unified solutions to various OCR tasks by taking advantage of some of the merits of LVLMs. These OCR specialist models usually adopt Transformer encoder-decoder architecture.

1.3.1 Traditional Methods

Traditional methods for text recognition usually adopt the framework of character segmentation and recognition. Characters are explicitly segmented for each text line after text line detection or segmentation. The features of the segmented characters are extracted and then fed into a statistical classifier for character recognition.

Character segmentation methods [15] typically utilize techniques such as projection analysis, connected component analysis, contour analysis, morphological approaches [82], and character stroke analysis [127], etc.

Commonly used feature extraction methods include directional features [75, 134], Gabor features [16, 32], scale-invariant feature transform (SIFT) [21, 68, 135] etc.

For character classification, the modified quadratic discriminant function (MQDF) [55, 110], support vector machine (SVM) [93, 103, 107, 109], K-nearest neighbors (KNN) [3, 5], and random forest [14, 127] algorithms are often adopted as classifiers. Finally, the recognition results can also be postprocessed by using a language model [79, 116].

The main disadvantage of character segmentation-based methods is that character segmentation errors result in limited performance of the subsequent character recognition. Accurately segmenting touched characters, which are common in handwritten text images, is difficult.

Unlike segmenting each character, sequence modeling-based methods attempt to recognize text line images by exploiting contextual dependencies in features, such as the hidden Markov model (HMM) [2, 28]. In addition, some methods also use feature matching to recognize word images holistically [4, 34, 91, 92]. However, these methods have limited model generalization capabilities and poor support for out-of-vocabulary words.

1.3.2 Deep Learning-Based Methods

The deep learning-based text recognition methods including CTC-based methods and encoder-decoder-based methods, are summarized in Table 1.1.

1.3.2.1 CTC-Based Text Recognition Methods

CTC [36] was originally proposed for speech recognition tasks to decode 1D time sequences. Since an image is two-dimensional, existing methods usually convert a text image into a 1D time sequence by sliding a window along the reading direction. For example, each element of a 1D sequence is a column of image pixels or a feature vector extracted from the image in the sliding window. CNN and recurrent neural networks (RNNs) with gating mechanism [22, 46] are usually used to extract spatial and temporal features. CTC-based models map each element of the feature sequence into a character or a "blank" symbol. The "blank" symbol is introduced to address cases where multiple identical characters are recognized consecutively. In this way, the network can be trained in an end-to-end manner without acquiring character-level annotations.

Table 1.1 Text recognition based on deep learning

Representation learning	Decoding	Literature
CNN (segmentation)	character classifier	[50, 61, 111, 136]
MDRNN	CTC	[35, 38, 118]
CNN	CTC	[31, 33, 57, 128]
CNN + RNN	CTC	[43, 86, 95]
CNN	attention	[9, 137]
CNN + RNN	attention	[71, 96, 98, 101, 124, 133]
CNN + self-attention network	attention	[94]
CNN	2D attention	[112]
CNN + RNN	2D attention	[58, 89]
CNN + self-attention network	2D attention	[74]
CNN + self-attention network	2D attention + language model	[29, 129]
CNN + self-attention network	mask-guided Transformer decoder	[60]
VitDet [59]	Qwen 0.5 B [10]	[115]

The CTC decoding method is widely used in handwriting recognition tasks, including Chinese [118], English [12, 13, 24, 86], French [13, 78, 86], Arabic [35], and scene text recognition tasks [43, 47, 63, 95, 100].

In the training stage, the objective function of the CTC is defined as

$$\mathcal{L}_{\text{CTC}} = -\log \sum_{\pi \in \mathcal{B}^{-1}(y)} p(\pi|x) \quad (1.1)$$

where x denotes the input sequence for the CTC, and where y denotes the ground-truth text. π denotes a decoding path output by the network, i.e., the per-frame or per-timestep predictions. $\mathcal{B}^{-1}(y)$ refers to all decoding paths that can generate the label y. In the test stage, the model first merges repeated characters between adjacent "blank" symbols and then removes all the "blank" symbols. Furthermore, the CTC also uses the beam search algorithm [117] for faster and more efficient decoding.

For the encoding method, Graves et al. first proposed a multidimensional recurrent neural network (MDRNN) [37] to convert a 2D image into a 1D feature sequence, which is convenient for CTC-based decoding. Models with an MDRNN-CTC framework are applied to tasks such as Arabic handwriting recognition [35] and Chinese handwriting recognition [118]. However, the MDRNN has a limited ability to represent visual information in images and has high computational complexity.

Instead of using the MDRNN, fully convolutional networks are also commonly adopted to extract feature maps [31, 33, 128]. An input image is gradually collapsed into a 1D feature sequence by using a carefully designed CNN for feature extraction. Compared with the MDRNN, the CNN can learn rich local structural information from images. For example, Yin et al. [128] used a sliding window on input images. The features of each window are extracted by a CNN and are composed of a feature

1.3 Multilingual Text Recognition Methods

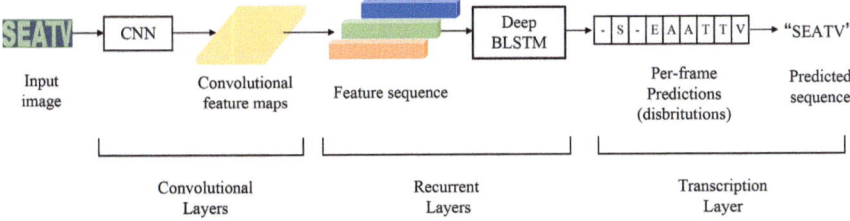

Fig. 1.2 Network structure of the CRNN

sequence. Wang et al. [108] proposed a gated recurrent convolutional network, which introduces a gating mechanism in a CNN. In these methods, the CTC is used for decoding.

Recent research has shown that network architectures that integrate both CNN and RNN outperform those that utilize only a single MDRNN or CNN [86, 95]. A typical network framework named CRNN, as shown in Fig. 1.2, uses a CNN and an RNN for feature extraction and encoding and uses a CTC for decoding. In this network framework, a VGG network [99] is used to extract feature sequences, and a bidirectional long short-term memory (LSTM) network [46] is used to learn contextual dependencies in the feature sequences.

1.3.2.2 Encoder-Decoder-Based Text Recognition Methods

The encoder-decoder-based model was originally applied in machine translation tasks [102]. Originally, RNNs were adopted as both the encoder and decoder in the model. The encoder aims to learn contextual dependencies from the feature sequence extracted by a CNN. The characters are then recursively decoded by the decoder. By introducing the attention mechanism [8] in the decoder, the model calculates correlation coefficients as attention weights between the current decoding character and the elements in the encoder's output feature sequence, thus learning to align the feature sequence and decoded characters.

Compared with CTC-based models, encoder-decoder models based on an attention mechanism learn an implicit language model in the decoding process, which can integrate more language priors. Therefore, encoder-decoder models with attention mechanism can often achieve higher accuracy than CTC-based models in scene text recognition tasks [67]. In recent years, language models that are based on Transformer encoders [105], such as BERT [23], XLNet [125] and RoBERTa [65], have shown effective performance in the NLP research field. Some studies [29, 129] propose the use of an additional language model based on Transformer encoder to revise the original predictions by incorporating semantic reasoning.

Encoder-decoder models based on attention mechanism are widely used in handwriting recognition [25, 73, 101, 113, 137] and scene text recognition [9, 19,

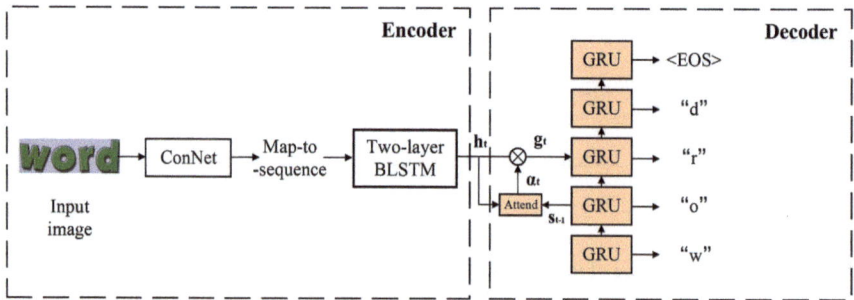

Fig. 1.3 An encoder-decoder model with attention mechanism

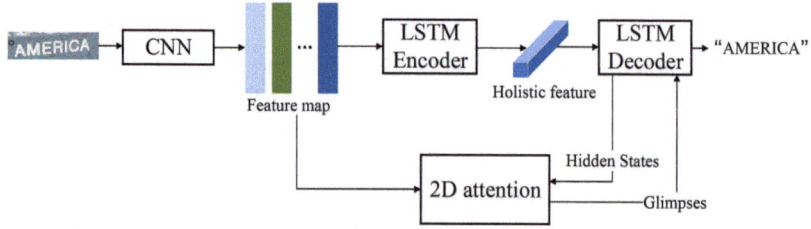

Fig. 1.4 A text recognition model based on 2D attention mechanism

29, 57, 62, 64, 87, 88, 94, 96, 98, 113, 129, 130, 138] tasks. Figure 1.3 shows an encoder-decoder model based on attention mechanism [96].

For the encoding method, similar to CTC-based methods, most encoder-decoder methods based on attention mechanisms use CNNs and RNNs to extract feature sequences from input images. Qin et al. [89] proved that directly flattening 2D feature maps into 1D and using an RNN for decoding can achieve high accuracy. However, such methods have limited performance on irregular text images with complex text layouts.

Attempts have been made to retain more 2D spatial information in features by improving the representation learning method. Cheng et al. [20] propose to encode the input image from four directions (top-to-bottom, bottom-to-top, left-to-right and right-to-left), and the features of different directions are weighted by weights learned by the model. However, this method can focus only on limited directional information, while the text layouts in real samples may be more complicated.

The 2D attention mechanism [123] is another promising solution. The SAR network [58] first applies the 2D attention mechanism in the text recognition task and significantly improves the recognition accuracy of irregular text images. As shown in Fig. 1.4, instead of being transformed into a feature sequence, the CNN in the model extracts 2D feature maps. During the decoding process, attention weights are calculated on the entire 2D feature map. Lyu et al. [74] proposed applying a Transformer encoder to model dependencies among elements in 2D feature maps. In

1.3 Multilingual Text Recognition Methods

GT: that one of us must volunteer . The Bishop , Dr. Burge , did

Pred: that one of us must volunteer the Bishop . The Bishop . Dr__uge , did

Fig. 1.5 An example of the attention drift problem

addition to exploiting 2D spatial information in the encoding stage, several methods have been proposed [106, 121] to decode along multiple directions.

Although models with 2D attention mechanisms have achieved high accuracy, they still have the following two problems.

First, attention weights are calculated on the entire 2D feature map, which increases the computational complexity. The recursive decoding process results in additional time costs. Therefore, it is difficult to apply models with 2D attention mechanism in practical applications that require high efficiency. For example, the lightweight OCR system (PP-OCR) [26, 27] developed by Baidu still adopts the "CNN-LSTM-CTC" framework.

Second, some studies [19, 113] have shown that attention-based encoder-decoder models suffer from the attention drift problem. Due to the absence of global visual information in the decoder inputs, the decoding process for attention-based encoder-decoder models is highly sensitive to the previously decoded results. As shown in Fig. 1.5, misalignments between features and characters lead to repeated characters or missing characters. In Fig. 1.5, arrows in the heatmap of attention weights denote the occurrence of the attention drift problem. Wrongly recognized characters are marked in red, and "_" refers to missing characters. The attention drift problem often occurs in images with long text, e.g., handwritten text images.

In summary, encoder-decoder models based on attention mechanism can learn alignments between features and text during decoding. The learned implicit language model helps exploit semantic information in text. In recent years, 2D attention mechanism has been proposed to recognize irregular text images, and the recognition accuracy has significantly improved. However, encoder-decoder models based on attention mechanism also suffer from high computational complexity and the attention drift problem.

Training deep learning models generally relies on the availability of large-scale datasets. Table 1.2 lists the information of the text recognition datasets used in this book. Detailed information can be found in the appendix.

Table 1.2 Text recognition datasets used in this book

Task	Dataset	Type	Language	Text length	#Train	#Val	#Test
Scene text	MJSynth	Synthetic	English	Word	8.9 M	–	–
	SynthText	Synthetic	English	Word	7.2 M	–	–
	IIIT5K	Real	English	Word	2000	–	3000
	SVT	Real	English	Word	257	–	647
	IC03	Real	English	Word	1156	–	867
	IC13	Real	English	Word	848	–	857
	IC15	Real	English	Word	4468	–	1811
	SVTP	Real	English	Word	–	–	645
	CUTE	Real	English	Word	–	–	288
	SynthChinese	Synthetic	Chinese + English	Word	1.0 M	–	–
	RCTW	Real	Chinese + English	Word	44,001	–	–
	MLT19	Real	Multilingual	Word	105,432	–	–
	AcTiV	Real	Arabic	Sentence	7943	814	930
Handwritten text	IAM	Real	English	Sentence	6161	966	2915
	IAM-word	Real	English	Word	53,839	16,465	17,616
	Rimes	Real	French	Sentence	10,171	1162	778
	IFN/ENIT	Real	Arabic	Word	26,459	–	6033

As shown in Table 1.2, most scene text images are segmented by words, whereas handwritten text images can often be segmented by text lines, which results in different data distributions between scene text images and handwritten text images.

1. **Text layouts**. Scene texts have arbitrary layouts, including horizontal, vertical, skewed and curved, as shown in Fig. 1.7b. Handwritten texts are usually arranged regularly.
2. **Text lengths**. Most scene text images are word-level, whereas handwritten text images contain many long texts. Figure 1.6 shows the distributions of the text lengths of the scene text datasets and handwriting datasets.

1.4 Datasets

1.4.1 Scene Text Datasets

It is difficult to collect sufficient real text images with annotations for training deep models [6]. Thus, most existing methods use synthetic samples for training, and the performance is evaluated on real samples.

1.4 Datasets

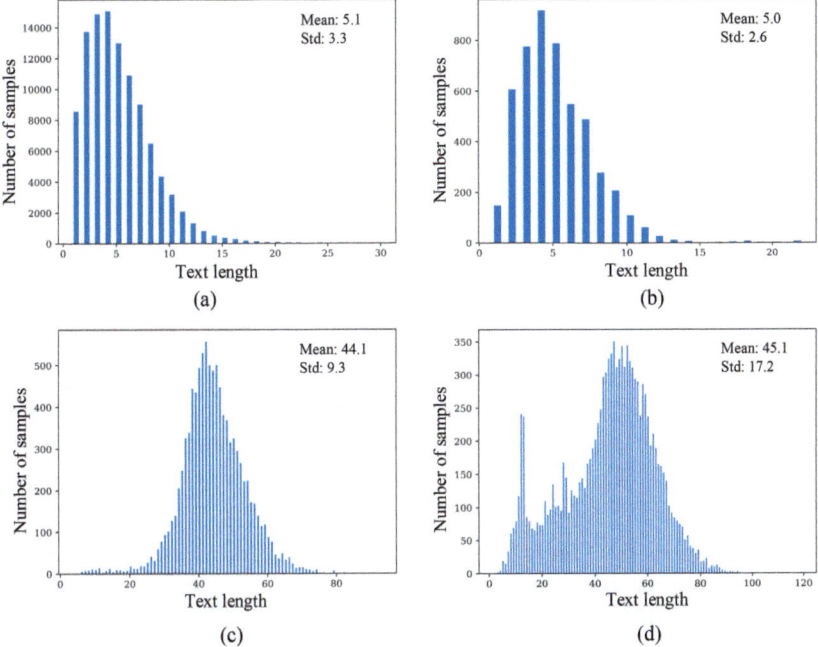

Fig. 1.6 Text length statistics of different datasets. (**a**) MLT19 multi-lingual scene text dataset. (**b**) IIIT5K English scene text dataset. (**c**) IAM English handwriting dataset. (**d**) Rimes French handwriting dataset

The following three synthetic scene text datasets are used in the experiments of this book.

1. **MJSynth** [49] contains approximately 8.9 million word images. The synthesis process can be divided into seven steps: (1) font rendering, (2) border and shadow rendering, (3) background coloring, (4) composition of the font, border and background, (5) text shape transformation, (6) blending with real images, and (7) mixing with noise. The samples in the MJSynth dataset are usually high-quality.
2. **SynthText** [40] is an end-to-end scene text detection and recognition dataset. The synthesis engine first finds possible text regions in a background image via depth estimation and then renders text in these regions. Usually, word-level images are cropped from original images according to bounding box annotations when training a scene text recognition model. There are approximately 7.2 million cropped word image samples. The SynthText dataset has considerable low-quality images.
3. **SynthChinese**: To train Chinese scene text recognition models, a SynthChinese dataset is established in our research work by collecting Chinese text corpora and fonts. The texts for the synthesis are selected from the THUOCL corpus

Fig. 1.7 Examples of regular and irregular text images. (**a**) Regular text images. (**b**) Irregular text images

[41]. Images are synthesized via the method proposed by Gupta et al. [40]. There are approximately 1 million cropped word image samples.

There are seven English scene text datasets derived from real application scenarios that are commonly used to evaluate model performance. These datasets can be divided into regular datasets and irregular datasets [6, 7]. Most samples in regular datasets are regularly arranged horizontal text images, whereas irregular datasets contain many skewed and curved text images. Figure 1.7 shows some examples of regular and irregular text images.

1. **IIIT5K-Words (IIIT5K)** [79] is a regular scene text dataset that contains 2000 training samples and 3000 testing samples.
2. **Street View Text (SVT)** [109] is a regular scene text dataset that contains 257 training samples and 647 testing samples.
3. **ICDAR2003 (IC03)** [70] is a regular scene text dataset that contains 1156 training samples and 1110 testing samples. According to the test protocol used by Cheng et al. [19], samples with text that is too short (less than 3 characters) or nonalphanumeric characters are ignored, and the remaining 867 samples are used as the test set in the experiments of this book.
4. **ICDAR2013 (IC13)** [53] is a regular scene text dataset that contains 848 training samples and 1095 testing samples. Similar to IC03, which ignores samples with text that is too short (less than 3) or nonalphanumeric characters, the remaining 857 samples are used as the test set in the experiments of this book.
5. **ICDAR2015 (IC15)** [54] is an irregular scene text dataset that contains 4468 training samples and 2077 testing samples. There are many low-quality images in the dataset. After ignoring samples that contain nonalphanumeric characters and extremely low-quality samples [9, 19], the remaining 1811 samples are used as the test set in the experiments of this book.
6. **SVT Perspective (SVTP)** [85] is an irregular scene text dataset that contains 645 test samples. Many samples are affected by perspective distortion.
7. **CUTE80 (CUTE)** [90] is an irregular scene text dataset that contains 288 test samples. There are many curved text images in the dataset.

To evaluate different models on multilingual scene text images, the following datasets are also used in this book.

1. **MLT19** [81] is a multilingual scene text dataset provided by the International Conference on Document Analysis and Recognition (ICDAR 2019) robust

1.4 Datasets

Fig. 1.8 Multilingual scene text images in the MLT19 dataset

reading challenge for multilingual scene text detection and recognition (RRC-MLT-2019). The dataset involves 10 different languages, in which English, French, German and Italian belong to the Latin writing system; Chinese, Japanese and Korean belong to the Chinese writing system; Arabic belongs to the Arabic writing system; and Bengali and Hindi belong to the Indian writing system. The dataset consists of unconstrained images captured by mobile phones. There are many blurred and arbitrarily oriented images. The training set contains 10,000 samples for end-to-end text detection and recognition. After cropping the text regions from the original images, 105,432 word-level images remained. The annotations of the test set are not publicly available, but model performance on the test set can be evaluated online.[1] Some examples of word-level images in the MLT19 dataset are shown in Fig. 1.8.

2. **RCTW** [97] is an end-to-end scene text detection and recognition dataset that contains both Chinese and English text images. The dataset does not provide ground-truth text for the test set. Therefore, this book selects 5000 and 1000 word-level images from the original training set as the new training set and test set, respectively. To evaluate model performance on multioriented text images that are common for Chinese, the ratio of horizontal to vertical text images is controlled to 1:1 in both the training set and test set.

3. **AcTiV** [131, 132] is an Arabic news video caption dataset provided by the International Conference on Document Analysis and Recognition (ICDAR 2017) and the International Conference on Pattern Recognition (ICPR 2020) competition on Arabic text detection and recognition in multiresolution video frames. The training set, validation set, and test set contain 7943, 814 and 930 samples, respectively.

1.4.2 Handwriting Datasets

Three public handwriting datasets are used in this book, i.e., the IAM English handwriting dataset IAM [77], the Rimes French handwriting dataset [39], and the

---

[1] https://rrc.cvc.uab.es/?ch=15

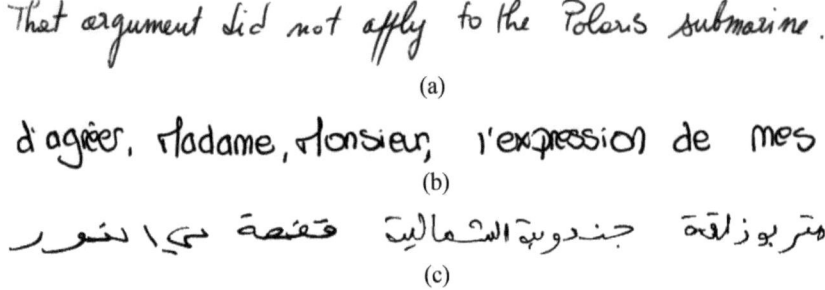

Fig. 1.9 Examples of handwritten text images. (**a**) Handwritten English image. (**b**) Handwritten French image. (**c**) Handwritten Arabic image

IFN/ENIT Arabic handwriting dataset [83]. Examples from the three datasets are shown in Fig. 1.9.

1. **IAM** [77] is an English handwriting dataset. The original data are collected from 1539 handwritten table images provided by 657 volunteers, and the text comes from the LOB corpus [51]. After segmenting the original images by text lines, there are 6161 training samples, 966 validation samples and 2915 test samples. The IAM dataset also provides word-level text images, including 53,839 training samples, 16,465 validation samples, and 17,616 test samples.
2. **Rimes** [39] is a French handwriting dataset that contains 10,171 training samples, 1162 validation samples and 778 test samples.
3. **IFN/ENIT** [83] is an Arabic handwriting dataset. The publicly available data include five subsets (a, b, c, d, e). One commonly used test scenario is abcd-e, i.e., the training set consists of a set through set d, and the test set is set e. Under this partition, the training set contains 26,459 samples, and the test set contains 6033 samples.

1.5 Evaluation Metrics

1.5.1 Evaluation Metrics for Text Recognition

For text recognition, different metrics are used for word-level and sentence-level text recognition.

For recognition of word-level images, word accuracy is usually used. It is defined as the proportion of correctly recognized samples among the total samples.

For the recognition of sentence-level images, it is more reasonable to use CER and WER. Both CER and WER are based on the edit distance between the prediction and the ground-truth text. The CER is calculated as the ratio of the character-level edit distance to the total number of characters in the ground-truth text, whereas the

1.5 Evaluation Metrics

WER is calculated as the ratio of the word-level edit distance to the total number of words in the ground-truth text. The CER and WER can be calculated as

$$\text{CER} = \frac{S_c + D_c + I_c}{N_c} \qquad (1.2)$$

$$\text{WER} = \frac{S_w + D_w + I_w}{N_w} \qquad (1.3)$$

where S_c, D_c and I_c refer to the numbers of replacements, deletions, and insertions at the character level that are required to recover the ground-truth text from the prediction. N_c is the number of characters in the ground-truth text. S_w, D_w and I_w refer to the numbers of replacements, deletions, and insertions at the word level that are required to recover the ground-truth text from the prediction. N_w is the number of words in the ground-truth text. The smaller the CER and WER are, the higher the model performance.

The character recognition rate (CRR) is calculated as follows:

$$CRR = 1 - CER \qquad (1.4)$$

In some cases, line recognition rate (LRR) is also used, which is defined as the ratio of the number of correctly recognized text line images to the total number of text line images in the dataset.

1.5.2 Evaluation Metrics for End-to-End Text Detection and Recognition

For end-to-end text detection and recognition, the evaluation metrics include precision, recall, F1 score, average precision and character recognition rate. A word is correctly predicted only if it is both correctly detected and recognized. The criterion of correct detection is that the intersection over union (IoU) between the label bounding box and the predicted results is greater than 0.5, and the criterion of correct recognition is that the prediction is the same as the ground-truth text in a case-insensitive manner. Precision is the ratio of the number of correctly detected and recognized text instances to the number of detected text instances. Recall is the ratio of the number of correctly detected and recognized text instances to the number of text instances in the ground truth. The F1 score is calculated as:

$$F_1 = 2 \cdot \frac{\text{precision} \cdot \text{recall}}{\text{precision} + \text{recall}} \qquad (1.5)$$

1.6 Book Overview

1.6.1 Problem Analysis

As reviewed in Sects. 1.3.1 and 1.3.2, although text recognition methods based on deep learning have achieved promising results in recent years, it is still challenging for existing methods to recognize text images taken in unconstrained environments accurately and efficiently. The key challenges include: (1) achieving both high recognition performance and efficiency in text recognition models, (2) learning complex spatial information in text images, and (3) modeling long-term contextual information in text images.

Achieving Both High Recognition Performance and Efficiency in Text Recognition Models Numerous practical applications, including autonomous driving and robot navigation, demand models that exhibit both high recognition accuracy and rapid recognition speed. However, CTC-based models have limited performance in irregular text images. Models with 2D attention mechanism have achieved higher accuracy on irregular text images, but the time costs of these models are generally high. Therefore, novel text recognition methods with high performance and high efficiency must be explored.

Learning Complex Spatial Information in Text Images Scene text images often have complex 2D structures, which introduce multilevel spatial dependencies in images. Research on human visual mechanism has shown that human visual perception includes scene analysis from local to global locations [45, 80], and this mechanism is applicable to the text image recognition process. Existing methods typically use RNNs or self-attention networks to learn spatial dependencies. For RNNs, 2D feature maps are transformed into 1D feature sequences as inputs, resulting in the loss of 2D spatial information. For self-attention networks, contextual information is better utilized, but there is still room for improvement in their ability to learn multiscale spatial dependencies. It is important to explore efficient ways to exploit 2D spatial information.

Modeling Long-Term Contextual Information in Text Images For images with long texts, it is crucial to learn long-term contextual information. In the encoding stage, existing methods generally adopt RNNs to learn contextual information in the feature sequence. An unfolded RNN has the same network depth as the length of the feature sequence after being expanded in the temporal dimension, which makes it difficult to train [11]. In the decoding stage, CTC-based methods cannot utilize the semantic information in texts, whereas encoder-decoder approaches with attention mechanism suffer from the attention drift problem. Therefore, both the encoding and decoding methods in existing text recognition models are still insufficient for handling images with long texts.

1.6.2 Research Content

To address the three main challenges mentioned above, the research content of this book can be divided into three parts, including primitive representation learning, multielement attention mechanism, and dynamic temporal residual learning with attention rectification. On the basis of the main research progress, a text recognition system framework named TH-DL is established for text detection and recognition in practical applications.

1.6.2.1 Primitive Representation Learning

Recognizing text in images is a basic function of human vision. Marr's vision computing theory [76] assumes that vision is a complex multistage representation and processing task. Multiple visual primitives can be used for different stages. For example, the primitives of a primal sketch include zero-crossings, blobs, and edge segments.

Inspired by Marr's visual computing theory, a primitive representation learning method is proposed in this book to find suitable primitive representations for text images. The primitive representations are learned via feature aggregation from feature maps extracted by a CNN. Two aggregation methods are explored: pooling aggregation, which adopts global average pooling, and weighted aggregation, which uses a spatial attention mechanism. Visual text representations can be obtained by combining the primitive representations, and each visual text representation corresponds to a character to be recognized. Therefore, primitive representations can be regarded as a series of base vectors for visual text representation.

The combination of primitive representations is modeled as a graph representation learning task, and is implemented by a graph convolutional network (GCN) [17, 56]. In this way, the implemented primitive representation learning network (PREN) model can generate visual text representations of all characters in parallel.

A semantic-guided decoding method is further proposed to exploit semantic information in the decoding process. The visual text representations are transformed into semantic representations via a language model that is based on a Transformer encoder [105]. The semantic representations are used to reweight primitive representations, which integrate both visual and semantic information.

1.6.2.2 Multielement Attention Mechanism

A multielement attention (MEA) mechanism is proposed in this book. Each element in the feature maps extracted by a CNN is modeled as a node of an undirected graph. By designing different adjacency matrices to aggregate features of neighboring nodes before calculating attention weights, the MEA can learn different levels of spatial information. Specifically, three types of adjacency matrices are used to learn the local, neighboring and global geometric information of text images in the implemented multielement attention network (MEAN). The self-attention mechanism [105] can be viewed as a special case of MEA. Compared with the self-attention mechanism, the MEA with different adjacency matrices can better model complex spatial dependencies in text images.

An approach called PREN2D is also proposed to integrate PREN and MEAN. The visual text representations learned by PREN are combined with character embedding vectors as the input of the decoder in MEAN. In this way, visual text representations can provide global visual guidance for the decoding process.

1.6.2.3 Dynamic Temporal Residual Learning with Attention Rectification

A dynamic temporal residual learning method is proposed to improve the ability of RNN encoders to model contextual dependencies in feature sequences. Models with shortcut connections such as ResNet [42] demonstrated that residual learning can ease the training of deep feedforward networks. As an RNN can also be regarded as a deep network when unfolded along the temporal dimension, adding residual connections between temporally adjacent outputs of an LSTM network is feasible. Dynamic weights are further applied to residual connections to better model temporal dependencies in feature sequences. An additional LSTM network or self-attention network is used to learn the dynamic weights.

An attention rectification method is also proposed to alleviate the attention drift problem for the attentional decoder. An attention rectification module based on a parallel attention mechanism is introduced into the original encoder-decoder model. The attention rectification module calculates attention weight correction amounts to rectify the original attention weights in the decoder. The CTC criterion [36] is used in the attention rectification module to enhance the character position information in the features.

Finally, the dynamic temporal residual learning and attention rectification methods can be integrated to further improve performance.

1.6.2.4 TH-DL Text Recognition System Framework

The TH-DL text recognition system framework is further designed and implemented, which integrates our text detection model [119] on the basis of a modified Mask R-CNN [44]. For irregular scene text recognition, the proposed PREN2D method is used. The developed multilingual scene text recognition system based on the TH-DL framework has achieved the leading performance in end-to-end text multilingual scene text detection and recognition tasks on the public RRC-MLT-2019 [81] leaderboard. For long text instances such as video subtitles, the experimental results show that the CTC-decoding-based model is more applicable. Our previous CTC decoding-based Arabic video subtitle recognition system won

1.6 Book Overview

first place in both the ICDAR 2017 and ICPR 2020 competitions on Arabic text detection and recognition in multiresolution video frames [131, 132], and the further developed ARN-DTRN can achieve better performance.

1.6.3 Book Structure

The structure of this book is shown in Fig. 1.10. The remaining chapters of the book are briefly described as follows.

Chapter 2 presents the proposed primitive representation learning method, and implements the primitive representation learning network (PREN), which is different from the commonly used CTC-based methods and encoder-decoder methods.

Chapter 3 introduces the proposed multielement attention mechanism, and implements the multielement attention network (MEAN). By integrating primitive representation learning into a model with a multielement attention mechanism, the PREN2D model is constructed, which can achieve better recognition performance.

Chapter 4 describes the proposed dynamic temporal residual learning and attention rectification methods and presents the dynamic temporal residual network (DTRN) and attention rectification network (ARN).

Chapter 5 compares the different models proposed in the previous chapters and presents the TH-DL multilingual text recognition system framework. Finally, conclusions are drawn.

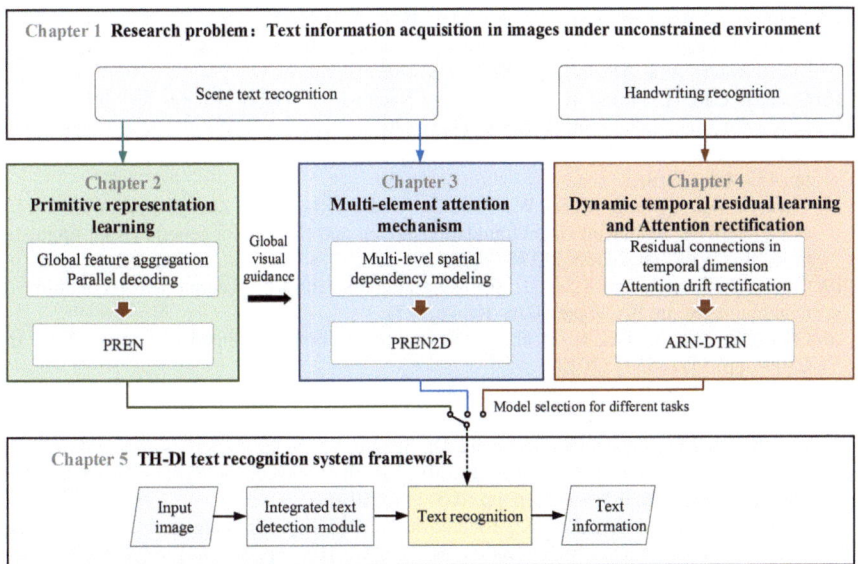

Fig. 1.10 The structure of this book

References

1. Achiam, J., Adler, S., Agarwal, S., Ahmad, L., Akkaya, I., Aleman, F.L., Almeida, D., Altenschmidt, J., Altman, S., Anadkat, S., et al.: GPT-4 technical report. arXiv preprint arXiv:2303.08774 (2023)
2. Ahmad, I., Rothacker, L., Fink, G.A., Mahmoud, S.A.: Novel sub-character HMM models for Arabic text recognition. In: ICDAR, pp. 658–662 (2013)
3. Alhutaish, R., Omar, N.: Arabic text classification using K-nearest neighbour algorithm. Int. Arab J. Inf. Technol. **12**(2), 190–195 (2015)
4. Almazán, J., Gordo, A., Fornés, A., Valveny, E.: Word spotting and recognition with embedded attributes. IEEE Trans. Pattern Anal. Mach. Intell. **36**(12), 2552–2566 (2014)
5. Altman, N.S.: An introduction to kernel and nearest-neighbor nonparametric regression. Am. Stat. **46**(3), 175–185 (1992)
6. Baek, J., Kim, G., Lee, J., et al.: What is wrong with scene text recognition model comparisons? Dataset and model analysis. In: ICCV, pp. 4715–4723 (2019)
7. Baek, J., Matsui, Y., Aizawa, K.: What if we only use real datasets for scene text recognition? Toward scene text recognition with fewer labels. In: CVPR, pp. 3113–3122 (2021)
8. Bahdanau, D., Cho, K., Bengio, Y.: Neural machine translation by jointly learning to align and translate. In: ICLR (2015)
9. Bai, F., Cheng, Z., Niu, Y., et al.: Edit probability for scene text recognition. In: CVPR, pp. 1508–1516 (2018)
10. Bai, J., Bai, S., Chu, Y., Cui, Z., Dang, K., Deng, X., Fan, Y., Ge, W., Han, Y., Huang, F., et al.: Qwen technical report. arXiv preprint arXiv:2309.16609 (2023)
11. Bengio, Y., Simard, P., Frasconi, P.: Learning long-term dependencies with gradient descent is difficult. IEEE Trans. Neural Netw. **5**(2), 157–166 (1994)
12. Bhunia, A.K., Das, A., Bhunia, A.K., Kishore, P.S.R., Roy, P.P.: Handwriting recognition in low-resource scripts using adversarial learning. In: CVPR, pp. 4767–4776 (2019)
13. Bluche, T.: Joint line segmentation and transcription for end-to-end handwritten paragraph recognition. In: NIPS, pp. 838–846 (2016)
14. Breiman, L.: Random forests. Mach. Learn. **45**(1), 5–32 (2001)
15. Casey, R., Lecolinet, E.: A survey of methods and strategies in character segmentation. IEEE Trans. Pattern Anal. Mach. Intell. **18**(7), 690–706 (1996)
16. Chen, J., Cao, H., Prasad, R., Bhardwaj, A., Natarajan, P.: Gabor features for offline Arabic handwriting recognition. In: DAS, pp. 53–58 (2010)
17. Chen, Y., Rohrbach, M., Yan, Z., et al.: Graph-based global reasoning networks. In: CVPR, pp. 433–442 (2019)
18. Chen, Z., Wu, J., Wang, W., Su, W., Chen, G., Xing, S., Zhong, M., Zhang, Q., Zhu, X., Lu, L., et al.: InternVL: scaling up vision foundation models and aligning for generic visual-linguistic tasks. In: CVPR, pp. 24185–24198 (2024)
19. Cheng, Z., Bai, F., Xu, Y., et al.: Focusing attention: towards accurate text recognition in natural images. In: ICCV, pp. 5076–5084 (2017)
20. Cheng, Z., Xu, Y., Bai, F., et al.: AON: towards arbitrarily-oriented text recognition. In: CVPR, pp. 5571–5579 (2018)
21. Chergui, L., Kef, M.: SIFT descriptors for Arabic handwriting recognition. Int. J. Comput. Vis. Robot. **5**(4), 441–461 (2015)
22. Cho, K., van Merriënboer, B., Bahdanau, D., Bengio, Y.: On the properties of neural machine translation: encoder–decoder approaches. In: Eighth Workshop on Syntax, Semantics and Structure in Statistical Translation, pp. 103–111 (2014)
23. Devlin, J., Chang, M.W., Lee, K., Toutanova, K.: BERT: pre-training of deep bidirectional transformers for language understanding. In: NAACL-HLT, pp. 4171–4186 (2019)
24. Ding, H., Chen, K., Yuan, Y., Cai, M., Sun, L., Liang, S., Huo, Q.: A compact CNN-DBLSTM based character model for offline handwriting recognition with Tucker decomposition. In: ICDAR, pp. 507–512 (2017)

References

25. Ding, H., Chen, K., Huo, Q.: An encoder-decoder approach to handwritten mathematical expression recognition with multi-head attention and stacked decoder. In: ICDAR, pp. 602–616 (2021)
26. Du, Y., Li, C., Guo, R., Yin, X., Liu, W., Zhou, J., Bai, Y., Yu, Z., Yang, Y., Dang, Q., et al.: PP-OCR: a practical ultra lightweight OCR system. arXiv preprint arXiv:2009.09941 (2020)
27. Du, Y., Li, C., Guo, R., Cui, C., Liu, W., Zhou, J., Lu, B., Yang, Y., Liu, Q., Hu, X., et al.: PP-OCRv2: bag of tricks for ultra lightweight OCR system. arXiv preprint arXiv:2109.03144 (2021)
28. Espana-Boquera, S., Castro-Bleda, M.J., Gorbe-Moya, J., Zamora-Martinez, F.: Improving offline handwritten text recognition with hybrid HMM/ANN models. IEEE Trans. Pattern Anal. Mach. Intell. **33**(4), 767–779 (2010)
29. Fang, S., Xie, H., Wang, Y., Mao, Z., Zhang, Y.: Read like humans: autonomous, bidirectional and iterative language modeling for scene text recognition. In: CVPR, pp. 7098–7107 (2021)
30. Feng, X., Yao, H., Zhang, S.: Focal CTC loss for Chinese optical character recognition on unbalanced datasets. Complexity **2019**(1), 9345861 (2019)
31. Feng, W., Yin, F., Zhang, X.Y., He, W., Liu, C.L.: Residual dual scale scene text spotting by fusing bottom-up and top-down processing. Int. J. Comput. Vis. **129**(3), 619–637 (2021)
32. Fogel, I., Sagi, D.: Gabor filters as texture discriminator. Biol. Cybern. **61**(2), 103–113 (1989)
33. Gao, Y., Chen, Y., Wang, J., et al.: Reading scene text with attention convolutional sequence modeling. Neurocomputing **339**, 161–170 (2019)
34. Gordo, A.: Supervised mid-level features for word image representation. In: CVPR, pp. 2956–2964 (2015)
35. Graves, A.: Offline Arabic handwriting recognition with multidimensional recurrent neural networks. In: Guide to OCR for Arabic scripts, pp. 297–313. Springer, London (2012)
36. Graves, A., Fernández, S., Gomez, F., Schmidhuber, J.: Connectionist temporal classification: labelling unsegmented sequence data with recurrent neural networks. In: ICML, pp. 369–376 (2006)
37. Graves, A., Fernández, S., Schmidhuber, J.: Multi-dimensional recurrent neural networks. In: ICANN, pp. 549–558 (2007)
38. Graves, A., Fernández, S., Liwicki, M., Bunke, H., Schmidhuber, J.: Unconstrained online handwriting recognition with recurrent neural networks. In: NIPS, pp. 577–584 (2008)
39. Grosicki, E., Carré, M., Brodin, J.M., Geoffrois, E.: Results of the Rimes evaluation campaign for handwritten mail processing. In: ICDAR, pp. 941–945 (2009)
40. Gupta, A., Vedaldi, A., Zisserman, A.: Synthetic data for text localisation in natural images. In: CVPR, pp. 2315–2324 (2016)
41. Han, S., Zhang, Y., Ma, Y., et al.: THUOCL: Tsinghua open Chinese lexicon (2016). http://thuocl.thunlp.org/
42. He, K., Zhang, X., Ren, S., Sun, J.: Deep residual learning for image recognition. In: CVPR, pp. 770–778 (2016)
43. He, P., Huang, W., Qiao, Y., Loy, C.C., Tang, X.: Reading scene text in deep convolutional sequences. In: AAAI, pp. 3501–3508 (2016)
44. He, K., Gkioxari, G., Dollár, P., Girshick, R.: Mask R-CNN. In: ICCV, pp. 2961–2969 (2017)
45. Hegdé, J.: Time course of visual perception: coarse-to-fine processing and beyond. Prog. Neurobiol. **84**(4), 405–439 (2008)
46. Hochreiter, S., Schmidhuber, J.: Long short-term memory. Neural Comput. **9**(8), 1735–1780 (1997)
47. Hu, W., Cai, X., Hou, J., et al.: GTC: guided training of CTC towards efficient and accurate scene text recognition. In: AAAI, pp. 11005–11012 (2020)
48. Hurst, A., Lerer, A., Goucher, A.P., Perelman, A., Ramesh, A., Clark, A., Ostrow, A., Welihinda, A., Hayes, A., Radford, A., et al.: GPT-4o system card. arXiv preprint arXiv:2410.21276 (2024)
49. Jaderberg, M., Simonyan, K., Vedaldi, A., et al.: Synthetic data and artificial neural networks for natural scene text recognition. In: NIPS Workshop on Deep Learning (2014)

50. Jaderberg, M., Vedaldi, A., Zisserman, A.: Deep features for text spotting. In: ECCV, pp. 512–528 (2014)
51. Johansson, S., Leech, G.N., Goodluck, H.: Manual of information to accompany the Lancaster-Oslo/Bergen Corpus of British English, for use with digital computer. Department of English, University of Oslo (1978)
52. Kang, L., Riba, P., Rusiñol, M., Fornés, A., Villegas, M.: Pay attention to what you read: non-recurrent handwritten text-line recognition. Pattern Recognit. **129**, 108766 (2022)
53. Karatzas, D., Shafait, F., Uchida, S., et al.: ICDAR 2013 robust reading competition. In: ICDAR, pp. 1484–1493 (2013)
54. Karatzas, D., Gomez-Bigorda, L., Nicolaou, A., et al.: ICDAR 2015 competition on robust reading. In: ICDAR, pp. 1156–1160 (2015)
55. Kimura, F., Takashina, K., Tsuruoka, S., Miyake, Y.: Modified quadratic discriminant functions and the application to Chinese character recognition. IEEE Trans. Pattern Anal. Mach. Intell. **9**(1), 149–153 (1987)
56. Kipf, T.N., Welling, M.: Semi-supervised classification with graph convolutional networks. In: ICLR (2017)
57. Lee, C.Y., Osindero, S.: Recursive recurrent nets with attention modeling for OCR in the wild. In: CVPR, pp. 2231–2239 (2016)
58. Li, H., Wang, P., Shen, C., Zhang, G.: Show, attend and read: a simple and strong baseline for irregular text recognition. In: AAAI, pp. 8610–8617 (2019)
59. Li, Y., Mao, H., Girshick, R., He, K.: Exploring plain vision transformer backbones for object detection. In: ECCV, pp. 280–296 (2022)
60. Li, T., Wu, S., Wang, Z.: Mask guided selective context decoding for handwritten Chinese text recognition. In: ICASSP, pp. 1–5 (2023)
61. Liao, M., Zhang, J., Wan, Z., Xie, F., Liang, J., Lyu, P., Yao, C., Bai, X.: Scene text recognition from two-dimensional perspective. In: AAAI, pp. 8714–8721 (2019)
62. Litman, R., Anschel, O., Tsiper, S., et al.: SCATTER: selective context attentional scene text recognizer. In: CVPR, pp. 11962–11972 (2020)
63. Liu, W., Chen, C., Wong, K.Y.K., Su, Z., Han, J.: STAR-Net: a spatial attention residue network for scene text recognition. In: BMVC (2016)
64. Liu, Z., Li, Y., Ren, F., et al.: Squeezedtext: a real-time scene text recognition by binary convolutional encoder-decoder network. In: AAAI, pp. 7194–7201 (2018)
65. Liu, Y., Ott, M., Goyal, N., Du, J., Joshi, M., Chen, D., Levy, O., Lewis, M., Zettlemoyer, L., Stoyanov, V.: RoBERTa: a robustly optimized BERT pretraining approach. arXiv preprint arXiv:1907.11692 (2019)
66. Liu, Y., Yang, B., Liu, J., Li, Z., Ma, Z., Zhang, S., Bai, X.: Textmonkey: an OCR-free large multimodal model for understanding document. arXiv preprint arXiv:2403.04473 (2024)
67. Long, S., He, X., Yao, C.: Scene text detection and recognition: the deep learning era. Int. J. Comput. Vis. **129**(1), 161–184 (2020)
68. Lowe, D.G.: Distinctive image features from scale-invariant keypoints. Int. J. Comput. Vis. **60**(2), 91–110 (2004)
69. Lu, N., Yu, W., Qi, X., Chen, Y., Gong, P., Xiao, R., Bai, X.: Master: multi-aspect non-local network for scene text recognition. Pattern Recognit. **117**, 107980 (2021)
70. Lucas, S.M., Panaretos, A., Sosa, L., et al.: ICDAR 2003 robust reading competitions: entries, results, and future directions. IJDAR **7**(2), 105–122 (2005)
71. Luo, C., Jin, L., Sun, Z.: MORAN: a multi-object rectified attention network for scene text recognition. Pattern Recognit. **90**(1), 109–118 (2019)
72. Lv, T., Huang, Y., Chen, J., Zhao, Y., Jia, Y., Cui, L., Ma, S., Chang, Y., Huang, S., Wang, W., et al.: Kosmos-2.5: a multimodal literate model. arXiv preprint arXiv:2309.11419 (2023)
73. Ly, N.T., Nguyen, H.T., Nakagawa, M.: 2D self-attention convolutional recurrent network for offline handwritten text recognition. In: ICDAR, pp. 191–204 (2021)
74. Lyu, P., Yang, Z., Leng, X., et al.: 2D attentional irregular scene text recognizer. arXiv preprint arXiv:1906.05708 (2019)

75. Ma, S., Xia, Y., Zhu, X.: Recognition of handwritten Chinese characters based on fuzzy direction line element features. J. Tsinghua Univ. Nat. Sci. Ed. **37**(3), 42–45 (1997)
76. Marr, D.: Vision: A Computational Investigation into the Human Representation and Processing of Visual Information. W. H. Freeman, San Francisco (1982)
77. Marti, U.V., Bunke, H.: The IAM-database: an English sentence database for offline handwriting recognition. Int. J. Doc. Anal. Recognit. **5**(1), 39–46 (2002)
78. Menasri, F., Louradour, J., Bianne-Bernard, A.L., Kermorvant, C.: The A2iA French handwriting recognition system at the Rimes-ICDAR2011 competition. In: DRR, pp. 263–270 (2012)
79. Mishra, A., Alahari, K., Jawahar, C.: Scene text recognition using higher order language priors. In: BMVC (2012)
80. Navon, D.: Forest before trees: the precedence of global features in visual perception. Cognit. Psychol. **9**(3), 353–383 (1977)
81. Nayef, N., Patel, Y., Busta, M., Chowdhury, P.N., Karatzas, D., Khlif, W., Matas, J., Pal, U., Burie, J.C., Liu, C.l., et al.: ICDAR2019 robust reading challenge on multi-lingual scene text detection and recognition–RRC-MLT-2019. In: ICDAR, pp. 1582–1587 (2019)
82. Nomura, S., Yamanaka, K., Katai, O., Kawakami, H., Shiose, T.: A novel adaptive morphological approach for degraded character image segmentation. Pattern Recognit. **38**(11), 1961–1975 (2005)
83. Pechwitz, M., Maddouri, S.S., Märgner, V., Ellouze, N., Amiri, H., et al.: IFN/ENIT-database of handwritten Arabic words. In: CIFED, pp. 127–136 (2002)
84. Peng, L., Liu, C., Ding, X., Wang, H.: Multilingual document recognition research and its application in China. In: DIAL, pp. 126–132 (2006)
85. Phan, T.Q., Shivakumara, P., Tian, S., et al.: Recognizing text with perspective distortion in natural scenes. In: ICCV, pp. 569–576 (2013)
86. Puigcerver, J.: Are multidimensional recurrent layers really necessary for handwritten text recognition? In: ICDAR, pp. 67–72 (2017)
87. Qiao, Z., Zhou, Y., Yang, D., et al.: SEED: semantics enhanced encoder-decoder framework for scene text recognition. In: CVPR, pp. 13528–13537 (2020)
88. Qiao, Z., Zhou, Y., Wei, J., Wang, W., Zhang, Y., Jiang, N., Wang, H., Wang, W.: PIMNet: a parallel, iterative and mimicking network for scene text recognition. In: ACM MM, pp. 2046–2055 (2021)
89. Qin, S., Bissacco, A., Raptis, M., Fujii, Y., Xiao, Y.: Towards unconstrained end-to-end text spotting. In: ICCV, pp. 4704–4714 (2019)
90. Risnumawan, A., Shivakumara, P., Chan, C.S., et al.: A robust arbitrary text detection system for natural scene images. Expert Syst. Appl. **41**(18), 8027–8048 (2014)
91. Rodriguez-Serrano, J.A., Perronnin, F., Meylan, F.: Label embedding for text recognition. In: BMVC (2013)
92. Rodriguez-Serrano, J.A., Gordo, A., Perronnin, F.: Label embedding: a frugal baseline for text recognition. Int. J. Comput. Vis. **113**(3), 193–207 (2015)
93. Roy, P.P., Pal, U., Lladós, J., Delalandre, M.: Multi-oriented and multi-sized touching character segmentation using dynamic programming. In: ICDAR, pp. 11–15 (2009)
94. Sheng, F., Chen, Z., Xu, B.: NRTR: a no-recurrence sequence-to-sequence model for scene text recognition. In: ICDAR, pp. 781–786 (2019)
95. Shi, B., Bai, X., Yao, C.: An end-to-end trainable neural network for image-based sequence recognition and its application to scene text recognition. IEEE Trans. Pattern Anal. Mach. Intell. **39**(11), 2298–2304 (2016)
96. Shi, B., Wang, X., Lyu, P., et al.: Robust scene text recognition with automatic rectification. In: CVPR, pp. 4168–4176 (2016)
97. Shi, B., Yao, C., Liao, M., Yang, M., Xu, P., Cui, L., Belongie, S., Lu, S., Bai, X.: ICDAR2017 competition on reading Chinese text in the wild (RCTW-17). In: ICDAR, pp. 1429–1434 (2017)
98. Shi, B., Yang, M., Wang, X., et al.: ASTER: an attentional scene text recognizer with flexible rectification. IEEE Trans. Pattern Anal. Mach. Intell. **41**(9), 2035–2048 (2019)

99. Simonyan, K., Zisserman, A.: Very deep convolutional networks for large-scale image recognition. In: ICLR (2015)
100. Su, B., Lu, S.: Accurate scene text recognition based on recurrent neural network. In: ACCV, pp. 35–48 (2014)
101. Sueiras, J., Ruiz, V., Sanchez, A., Velez, J.F.: Offline continuous handwriting recognition using sequence to sequence neural networks. Neurocomputing **289**(C), 119–128 (2018)
102. Sutskever, I., Vinyals, O., Le, Q.V.: Sequence to sequence learning with neural networks. In: NIPS, pp. 3104–3112 (2014)
103. Suykens, J.A., Vandewalle, J.: Least squares support vector machine classifiers. Neural Process. Lett. **9**(3), 293–300 (1999)
104. Team, G., Anil, R., Borgeaud, S., Alayrac, J.B., Yu, J., Soricut, R., Schalkwyk, J., Dai, A.M., Hauth, A., Millican, K., et al.: Gemini: a family of highly capable multimodal models. arXiv preprint arXiv:2312.11805 (2023)
105. Vaswani, A., Shazeer, N., Parmar, N., et al.: Attention is all you need. In: NIPS, pp. 5998–6008 (2017)
106. Wan, Z., Xie, F., Liu, Y., Bai, X., Yao, C.: 2D-CTC for scene text recognition. arXiv preprint arXiv:1907.09705 (2019)
107. Wang, K., Belongie, S.: Word spotting in the wild. In: ECCV, pp. 591–604 (2010)
108. Wang, J., Hu, X.: Gated recurrent convolution neural network for OCR. In: NIPS, pp. 334–343 (2017)
109. Wang, K., Babenko, B., Belongie, S.: End-to-end scene text recognition. In: ICCV, pp. 1457–1464 (2011)
110. Wang, Q.F., Yin, F., Liu, C.L.: Improving handwritten Chinese text recognition by confidence transformation. In: ICDAR, pp. 518–522 (2011)
111. Wang, T., Wu, D.J., Coates, A., Ng, A.Y.: End-to-end text recognition with convolutional neural networks. In: ICPR, pp. 3304–3308 (2012)
112. Wang, P., Yang, L., Li, H., Deng, Y., Shen, C., Zhang, Y.: A simple and robust convolutional-attention network for irregular text recognition. arXiv preprint arXiv:1904.01375 (2019)
113. Wang, T., Zhu, Y., Jin, L., Luo, C., Chen, X., Wu, Y., Wang, Q., Cai, M.: Decoupled attention network for text recognition. In: AAAI, pp. 12216–12224 (2020)
114. Wang, P., Bai, S., Tan, S., Wang, S., Fan, Z., Bai, J., Chen, K., Liu, X., Wang, J., Ge, W., et al.: Qwen2-VL: enhancing vision-language model's perception of the world at any resolution. arXiv preprint arXiv:2409.12191 (2024)
115. Wei, H., Liu, C., Chen, J., Wang, J., Kong, L., Xu, Y., Ge, Z., Zhao, L., Sun, J., Peng, Y., et al.: General OCR theory: towards OCR-2.0 via a unified end-to-end model. arXiv preprint arXiv:2409.01704 (2024)
116. Weinman, J., Learned-Miller, E., Hanson, A.: Fast lexicon-based scene text recognition with sparse belief propagation. In: ICDAR, pp. 979–983 (2007)
117. Wiseman, S., Rush, A.M.: Sequence-to-sequence learning as beam-search optimization. In: EMNLP, pp. 1296–1306 (2016)
118. Wu, Y.C., Yin, F., Chen, Z., Liu, C.L.: Handwritten Chinese text recognition using separable multi-dimensional recurrent neural network. In: ICDAR, pp. 79–84 (2017)
119. Xiao, S., Peng, L., Yan, R., An, K., Yao, G., Min, J.: Sequential deformation for accurate scene text detection. In: ECCV, pp. 108–124 (2020)
120. Xiao, B., Wu, H., Xu, W., Dai, X., Hu, H., Lu, Y., Zeng, M., Liu, C., Yuan, L.: Florence-2: Advancing a unified representation for a variety of vision tasks. In: CVPR, pp. 4818–4829 (2024)
121. Xie, Z., Huang, Y., Zhu, Y., et al.: Aggregation cross-entropy for sequence recognition. In: CVPR, pp. 6538–6547 (2019)
122. Xie, X., Fu, L., Zhang, Z., Wang, Z., Bai, X.: Toward understanding wordart: corner-guided transformer for scene text recognition. In: ECCV, pp. 303–321 (2022)
123. Xu, K., Ba, J.L., Kiros, R., Cho, K., Courville, A., Salakhutdinov, R., Zemel, R.S., Bengio, Y.: Show, attend and tell: neural image caption generation with visual attention. In: ICML, pp. 2048–2057 (2015)

124. Yang, M., Guan, Y., Liao, M., et al.: Symmetry-constrained rectification network for scene text recognition. In: ICCV, pp. 9147–9156 (2019)
125. Yang, Z., Dai, Z., Yang, Y., Carbonell, J., Salakhutdinov, R.R., Le, Q.V.: XLNet: generalized autoregressive pretraining for language understanding. In: NeurIPS, pp. 5753–5763 (2019)
126. Yang, Z., Tang, J., Li, Z., Wang, P., Wan, J., Zhong, H., Liu, X., Yang, M., Wang, P., Liu, Y., et al.: CC-OCR: a comprehensive and challenging OCR benchmark for evaluating large multimodal models in literacy. arXiv preprint arXiv:2412.02210 (2024)
127. Yao, C., Bai, X., Shi, B., Liu, W.: Strokelets: a learned multi-scale representation for scene text recognition. In: CVPR, pp. 4042–4049 (2014)
128. Yin, F., Wu, Y.C., Zhang, X.Y., Liu, C.L.: Scene text recognition with sliding convolutional character models. arXiv preprint arXiv:1709.01727 (2017)
129. Yu, D., Li, X., Zhang, C., et al.: Towards accurate scene text recognition with semantic reasoning networks. In: CVPR, pp. 12113–12122 (2020)
130. Yue, X., Kuang, Z., Lin, C., et al.: RobustScanner: dynamically enhancing positional clues for robust text recognition. In: ECCV, pp. 135–151 (2020)
131. Zayene, O., Ingold, R., BenAmara, N.E., Hennebert, J.: ICDAR2017 competition on Arabic text detection and recognition in multi-resolution video frames. In: ICDAR, pp. 1460–1465 (2020)
132. Zayene, O., Ingold, R., BenAmara, N.E., Hennebert, J.: ICPR2020 competition on text detection and recognition in Arabic news video frames. In: ICPR, pp. 343–353 (2020)
133. Zhan, F., Lu, S.: ESIR: end-to-end scene text recognition via iterative image rectification. In: CVPR, pp. 2059–2068 (2019)
134. Zhang, R., Ding, X., Fang, C.: A new method of optimal sampling features for offline handwritten Chinese character recognition (in Chinese). Chin. J. Image Graph. 7(2), 176–180 (2002)
135. Zhang, Z., Jin, L., Ding, K., Gao, X.: Character-SIFT: a novel feature for offline handwritten Chinese character recognition. In: ICDAR, pp. 763–767 (2009)
136. Zhang, Z., Zhang, C., Shen, W., Yao, C., Liu, W., Bai, X.: Multi-oriented text detection with fully convolutional networks. In: CVPR, pp. 4159–4167 (2016)
137. Zhang, Y., Nie, S., Liu, W., Xu, X., Zhang, D., Shen, H.T.: Sequence-to-sequence domain adaptation network for robust text image recognition. In: CVPR, pp. 2740–2749 (2019)
138. Zhang, C., Xu, Y., Cheng, Z., Pu, S., Niu, Y., Wu, F., Zou, F.: SPIN: structure-preserving inner offset network for scene text recognition. In: AAAI, pp. 3305–3314 (2021)

Chapter 2
Primitive Representation Learning

Abstract Different from CTC-based methods and encoder-decoder-based methods, this chapter proposes a primitive representation learning method that uses global feature aggregation to learn primitive representations from text images. Primitive representations can be regarded as a set of basis vectors in the feature space. Different combinations of primitive representations can generate visual text representations corresponding to the characters to be recognized. Visual text representations can be used for parallel decoding in the implemented primitive representation learning network (PREN). PREN can support both horizontal and vertical text in natural scene images. A semantic-guided decoding method is further incorporated to improve model performance on low-quality images by exploiting both visual and semantic information.

Keywords Primitive representation learning · Global feature aggregation · Visual text representation · Parallel decoding

2.1 Introduction of Primitive Representation Learning

There are two main types of deep learning-based sequence modeling methods for text recognition tasks. The first is the CRNN framework [12, 14, 22, 35, 38] which encodes images into hidden representations via CNNs and RNNs and uses connectionist temporal classification (CTC) [9] for decoding, as shown in Fig. 2.1a. The second is the attention-based encoder-decoder framework [1, 2, 4, 19, 21, 23, 31, 34, 35, 43, 46], which recursively converts the hidden representations output by the encoder into the predicted text through the attention mechanism, as shown in Fig. 2.1b. In the decoding process, the calculated attention weights can reflect the alignments between extracted features and predicted characters.

However, for CRNN-based text recognition methods, input images are often collapsed into 1D feature sequences during feature extraction. This process destroys the 2D spatial structure of the characters in the image and introduces redundant background information into the features. As a result, the performance of CRNN-based methods on irregular text images is often limited. However, for the encoder-

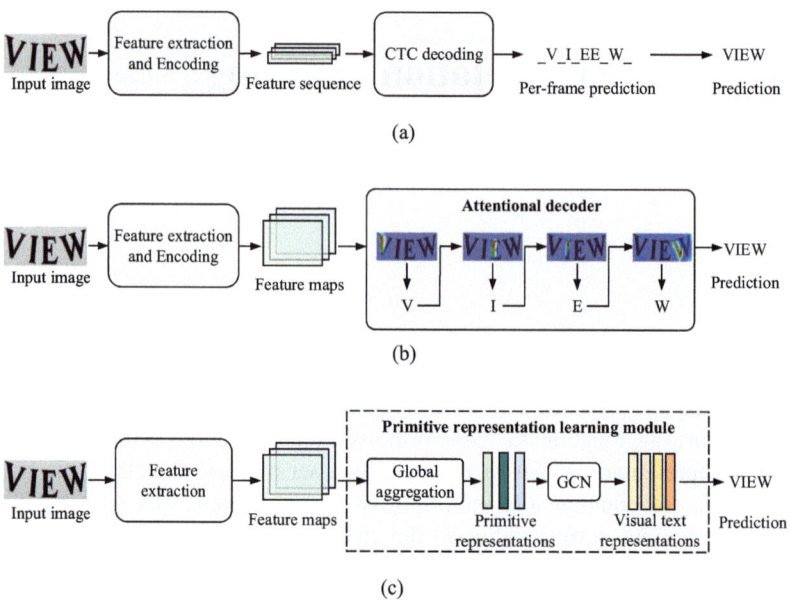

Fig. 2.1 Schematic diagram of different text image recognition methods. (**a**) CTC-based models. (**b**) Encoder-decoder models based on attention mechanism. (**c**) Primitive representation learning network

decoder model based on the attention mechanism, the decoder inputs consist of two parts: the character embedding vectors transformed from text labels in the training stage or decoded results in the inference stage, and the hidden representations output by the encoder. Among them, the character embedding vectors only contain semantic information and lack global visual information, which makes the decoding process sensitive to previous decoded results. When the contextual dependency is uncertain, it is difficult for the model to align decoded characters with corresponding regions in features and therefore suffers from the attention drift problem. Moreover, the attentional decoder often recursively predicts characters, increasing the time complexity of the model. On the basis of the above analysis, to better address complex fonts, backgrounds, and arrangements in text images, as well as to improve model recognition efficiency, we need to explore more efficient sequence modeling methods.

This chapter proposes a novel text recognition framework that uses global feature aggregation to learn primitive representations from text images. Primitive representations can be regarded as a set of basis vectors. Different combinations of primitive representations can generate visual text representations corresponding to the characters to be recognized. Visual text representations can be used for parallel decoding. Visual text representations are different from the character embedding vectors that are commonly used in encoder-decoder models. Visual text representations are transformed from primitive representations that contain visual

2.1 Introduction of Primitive Representation Learning

information of input images, whereas character embedding vectors are obtained from text with only semantic information.

For global feature aggregation, a pooling aggregator and a weighted aggregator are proposed. For the pooling aggregator, each primitive representation is learned from input feature maps through two convolutions followed by a global average pooling layer. In this way, aggregating weights are shared by all samples to learn intrinsic structural information from various scene text instances. For the weighted aggregator, a spatial attention mechanism is adopted to learn the importance of each position in the feature maps, which is used as aggregation weights. In this way, primitive representations can focus on the foreground of feature maps.

A primitive representation learning network (PREN) is further implemented, as shown in Fig. 2.1c. PREN contains a feature extraction module and a primitive representation learning module. The feature extraction module is used to extract 2D feature maps. In the process of feature extraction, multiscale features can be used to explore visual information with different receptive fields. The primitive representation learning module transforms feature maps into primitive representations through global feature aggregation and uses graph convolutional networks (GCNs) to generate visual text representations from primitive representations.

To make better use of the semantic information in the text, this chapter also combines the PREN with a language model and proposes a semantic-guided decoding method for the PREN. Inspired by pretrained language models [7, 24, 45] in the field of natural language processing (NLP) in recent years, this chapter adopts a Transformer encoder [41] as the language model, and uses the semantic representation output by the language model to reweight the primitive representations learned by the PREN. In this way, both visual information and semantic information can be utilized.

The performance of the PREN is evaluated on seven English scene text image datasets (IIIT5K [28], SVT [42], IC03 [27], IC13 [17], IC15 [18], SVTP [30] and CUTE [33]), RCTW Chinese scene text image dataset [36], and IFN/ENIT handwritten Arabic dataset [29]. The experimental results show that the PREN has both satisfactory recognition accuracy and recognition efficiency.

In summary, the main contributions of this chapter are as follows:

1. This chapter proposes a primitive representation learning method to learn efficient feature representations of the input image through global feature aggregation and realizes parallel decoding of text recognition by utilizing primitive representations.
2. This chapter proposes a pooling aggregation method and a weighted aggregation method to learn the primitive representations of input images and proposes the use of graph convolutional networks to transform primitive representations into visual text representations.
3. This chapter proposes a semantic-guided decoding method for the PREN, which uses a language model to convert visual text representations to semantic representations and uses semantic representations to reweight the primitive representations.

The rest of this chapter first describes the learning of primitive representations and visual text representations, then introduces the model structure of the PREN, as well as the semantic-guided decoding method that further combines the PREN with a language model, and finally provides the experimental results. The main content of this chapter comes from the authors' published paper [44], and the related code has been open-source.[1]

2.2 Primitive Representations

2.2.1 General Form of Primitive Representations

To retain sufficient global visual information in features, this chapter proposes performing global feature aggregation on the coordinate space of 2D feature maps extracted by a CNN to learn a set of primitive representations. Different combinations of primitive representations can generate visual text representations for each character.

Let $F \in \mathbb{R}^{d \times h \times w}$ be the feature maps output by a CNN, where h, w and d are the height, width and number of channels of F, respectively. The feature maps F are first converted into a feature matrix $X \in \mathbb{R}^{m_0 \times d}$, where $m_0 = h \times w$. Assuming that the number of primitive representations to be learned is n, the general form of primitive representations on the basis of global feature aggregations can be formulated as:

$$Z_i = f^{(i)}(X), \ i = 1, 2, \ldots, n \tag{2.1}$$

$$\boldsymbol{p}_i = \boldsymbol{a}_i \cdot Z_i, \ i = 1, 2, \ldots, n \tag{2.2}$$

where $\boldsymbol{p}_i \in \mathbb{R}^{1 \times d}$ is the i-th primitive representation, $f^{(i)}(\cdot)$ is the mapping function of a sub-network, $f^{(i)}(\cdot)$ converts the feature matrix X into hidden representations $Z_i \in \mathbb{R}^{m \times d}$, and $\boldsymbol{a}_i \in \mathbb{R}^{1 \times m}$ are the aggregation weights corresponding to the i-th primitive representation. The n primitive representations generated by the above process are concatenated to obtain the final primitive representations $P = [\boldsymbol{p}_1; \boldsymbol{p}_2; \ldots; \boldsymbol{p}_n] \in \mathbb{R}^{n \times d}$.

According to Eqs. (2.1) and (2.2), the i-th primitive representation is related to the mapping function $f^{(i)}(\cdot)$ and aggregation weights $\boldsymbol{a}_i$. By designing different mapping functions and aggregation weights, primitive representations can learn different spatial information.

If all the elements in the hidden representations share the same aggregating weight (i.e., $a_{ij} = \frac{1}{m}$, where m is the total number of elements in the hidden representations), primitive representations can learn intrinsic structural information

---
[1] https://github.com/RuijieJ/pren

2.2 Primitive Representations

from various scene text instances. Different mapping functions $f^{(i)}(\cdot)$ can be used for each primitive representation p_i to increase the diversity of primitive representations.

If all the elements in the hidden representations have different aggregating weights and the aggregating weights corresponding to each primitive representation p_i have a different distribution, then each primitive representation can focus on different regions in the input image. If the foreground region of the input image has higher aggregating weights, the primitive representation can reduce the redundant noise of the background region. Under this condition, all primitive representations can share the same mapping functions to reduce the computational complexity.

Two specific primitive representation learning schemes are designed, namely pooling aggregation and weighted aggregation. The two aggregation methods have different mapping functions $f^{(i)}(\cdot)$ and different aggregating weights a_i.

2.2.2 Pooling Aggregator

Figure 2.2 shows the structure of the pooling aggregator, where conv$_1$ and conv$_2$ are convolutional layers, #k denotes the number of convolutional kernels, d is the number of channels in the output feature maps, and n is the number of primitive representations.

The mapping function $f^{(i)}(\cdot)$ in Eq. (2.1) is implemented as two convolutional layers, i.e., "conv$_1$" and "conv$_2$" in Fig. 2.2. The input feature maps extracted by a CNN are converted into hidden representations Z_i, $i = 1, 2, \ldots, n$.

The pooling aggregator uses a global average pooling layer for feature aggregation, which is equivalent to setting $a_{ij} = \frac{1}{m}$, $\forall j = 1, 2, \ldots, m$ in Eq. (2.2), where a_{ij} is the aggregating weight of the j-th element in the hidden representations Z_i. Through global average pooling, different samples can share the same aggregating weights, which may help the model learn the global visual information of input images [13, 20].

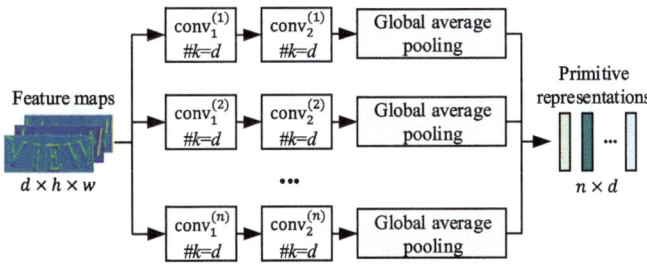

Fig. 2.2 Pooling aggregator

Combining the convolutional layers and the global average pooling operation, the calculation of the pooling aggregator can be formulated as

$$p_i = \text{Pool}\left(\text{conv}_2^{(i)}\left(\phi\left(\text{conv}_1^{(i)}(F)\right)\right)\right), \quad i = 1, 2, \ldots, n \qquad (2.3)$$

where $\text{conv}_1^{(i)}$ and $\text{conv}_2^{(i)}$, $i = 1, 2, \ldots, n$ are both convolutional layers with a kernel size of 3×3, stride of 2×2, and padding size of 1×1. $\phi(\cdot)$ denotes the Swish [32] activation function.

2.2.3 Weighted Aggregator

Owing to the diversity of text instances in natural scene images, it is also important to learn sample-specific information. Setting different aggregation weights for different positions in the feature map is beneficial for addressing irregular text images with complex text arrangements. Therefore, this chapter proposes the use of a spatial attention mechanism to learn aggregating weights from input features dynamically.

As shown in Fig. 2.3, the weighted aggregator uses a convolutional layer conv3 as the mapping function $f^{(i)}(\cdot)$ in Eq. (2.1), which converts the input feature maps F into hidden representations Z. Another convolutional layer conv4 and a sigmoid activation function transform the input feature maps F into heatmaps $H \in \mathbb{R}^{n \times h \times w}$. The i-th heatmap H_i corresponds to the i-th primitive representation, and the aggregating weights a_i can be calculated by flattening the i-th heatmap H_i. Primitive representations can be calculated via a scale-dot product and summation operation as follows.

$$Z = \phi\left(\text{conv}_3(F)\right) \qquad (2.4)$$

$$H = \sigma\left(\text{conv}_4(F)\right) \qquad (2.5)$$

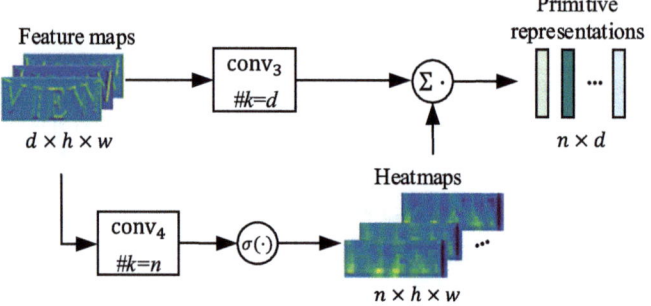

Fig. 2.3 Weighted aggregator

2.3 Visual Text Representations

$$p_i = \sum_{x=1}^{h}\sum_{y=1}^{w} H_{i,x,y} Z_{x,y} \qquad (2.6)$$

Equations (2.4)–(2.6) describe the calculation process of the weighted aggregator, where conv$_3$ and conv$_4$ are both convolutional layers with a kernel size of 3×3, stride of 1×1, and padding size of 1×1. $\sigma(\cdot)$ and $\phi(\cdot)$ represent sigmoid and Swish [32] activation functions, respectively. Equation (2.6) can be implemented by flattening the heatmap H in the spatial dimension to a matrix of size $\mathbb{R}^{n \times hw}$, flattening and transposing the hidden representations Z to a matrix of size $\mathbb{R}^{hw \times d}$, and computing the matrix product of the above two matrices. In this way, n primitive representations can be calculated in parallel.

2.3 Visual Text Representations

Since the primitive representations learned via global aggregation contain rich visual information of an input image, they can be transformed into features with textual information. A straightforward approach is to use linear combinations of primitive representations to generate visual text representations for characters. To enhance the nonlinear modeling ability, a fully connected layer with a nonlinear activation function can be used after the linear combination. On the basis of the above analysis, the process of generating visual text representations can be formulated as

$$V = \phi(APW) \qquad (2.7)$$

where $P \in \mathbb{R}^{n \times d}$ denotes n primitive representations, and d is the feature dimension. $A \in \mathbb{R}^{L \times n}$ is the coefficient matrix of the linear combination, and L is the preset maximum decoding length. $W \in \mathbb{R}^{d \times d}$ is the weight matrix of a fully connected layer, and $\phi(\cdot)$ is the Swish [32] activation function.

Equation (2.7) can be implemented as a GCN, where the coefficient matrix A of the linear combination is similar to an adjacency matrix, as shown in Fig. 2.4. The difference between visual text representation generation and regular GCNs is that the adjacency matrix of a regular GCN is usually square, whereas the size of the

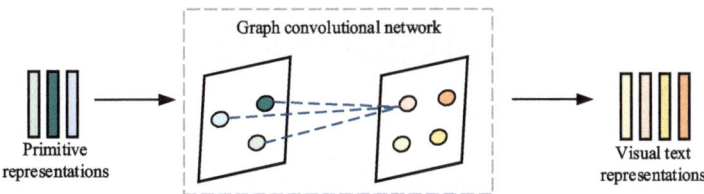

Fig. 2.4 Calculation process of visual text representations

coefficient matrix A is determined by the number of primitive representations n and the maximum decoding length L. Since the primitive representations do not have an explicit graph structure, the coefficient matrix A is randomly initialized and updated in the training stage.

Each visual text representation $\boldsymbol{v}_i$ ($i = 1, 2, \ldots, L$) can be used to represent a character to be recognized. For text with a length less than L, the padding character $\langle$pad$\rangle$ is used to extend the text length to L.

2.4 Primitive Representation Learning Network

The primitive representation learning network (PREN) consists of a feature extraction module and a primitive representation learning module. For the feature extraction module, different CNNs can be used, and multiscale feature maps can be extracted to make better use of visual information. The primitive representation learning module consists of global aggregators and a GCN. Finally, the PREN uses the visual text representations output by the primitive representation learning module for parallel decoding.

2.4.1 Feature Extraction

CNNs with good performance in general object recognition tasks can be adopted as backbone networks for feature extraction modules, such as ResNet [11], DenseNet [15], and EfficientNet [39]. This chapter uses two main types of CNN backbone networks: ResNet-50 and EfficientNet-B3. ResNet-50 has a faster recognition speed, whereas EfficientNet-B3 achieves higher recognition accuracy on the ImageNet dataset [6]. Figure 2.5 shows the structure of the two CNNs, where h and w are the height and width of the input image, respectively.

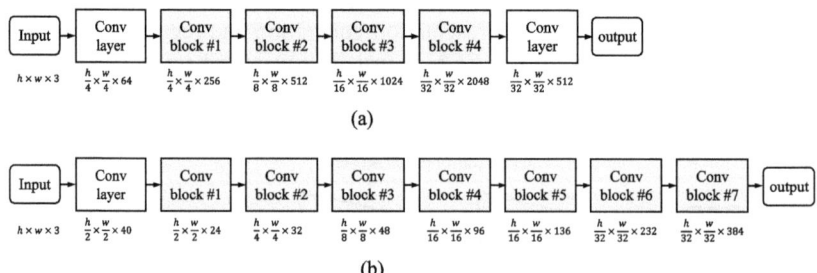

Fig. 2.5 The two CNN backbone networks used by the PREN. (**a**) ResNet-50. (**b**) EfficientNet-B3

ResNet-50 [11] contains a convolutional layer and four stacked convolutional blocks. Each convolutional block contains two 1×1 convolutional layers and one 3×3 convolutional layer, and there is a residual connection between the input and output of each convolutional block. The original ResNet network for image classification uses average pooling and fully connected layers to convert the output feature maps into a vector for classification. Since PREN needs only the feature maps output by the CNN rather than image classification, the average pooling and fully connected layers are removed. As shown in Fig. 2.5a, after each convolutional block, both the width and the height of the feature maps are reduced by half, and the number of channels is expanded by two or four times. The final feature maps of the original ResNet-50 model have 2048 channels. To reduce the computational complexity of the subsequent primitive representation learning module, an additional 1×1 convolutional layer is added to convert the channel number into 512.

EfficientNet-B3 [39] contains a convolutional layer and seven stacked convolutional blocks. Like ResNet-50, this chapter also removes the original pooling layer and fully connected layer to output 2D feature maps. The convolutional block in EfficientNet-B3 is the mobile inverted bottleneck (MBConv) [40] with a squeeze-and-excitation module. The mobile inversion bottleneck uses depthwise separable convolutions to reduce the computational complexity, and the squeeze-and-excitation module introduces attention mechanism in the channel dimension of feature maps. As shown in Fig. 2.5b, the spatial sizes of the feature maps output by the CNN are $\frac{1}{2}$, $\frac{1}{4}$, $\frac{1}{8}$, $\frac{1}{16}$ and $\frac{1}{32}$, respectively, of the original input image.

For tasks with high efficiency requirements, the PREN can directly use feature map output by the final convolutional layer for primitive representation learning. Because the spatial size of the feature maps is only $\frac{1}{32}$ of the input image, the computational complexity of the PREN is lower than that of the CTC-based models and the attention-based encoder-decoder. CTC-based models need to ensure that the length of the feature sequence is greater than the text length, whereas the attention-based encoder-decoder models follow a recursive decoding process.

For tasks with high accuracy requirements, the PREN can further use multiscale feature maps for primitive representation learning to better exploit structural information. Figure 2.6 shows the network structure of the PREN, which uses EfficientNet-B3 as the backbone network of the feature extraction module and extracts multiscale feature maps (1/8, 1/16, and 1/32 of the input image scale).

2.4.2 Primitive Representation Learning

For feature maps output by each selected convolutional block, the PREN uses a pooling aggregator and a weighting aggregator to learn primitive representations. For the case of using multiscale feature maps, the PREN unifies the dimension of each primitive representation as the number of channels of feature maps output by the final convolutional layer. For example, for the PREN shown in

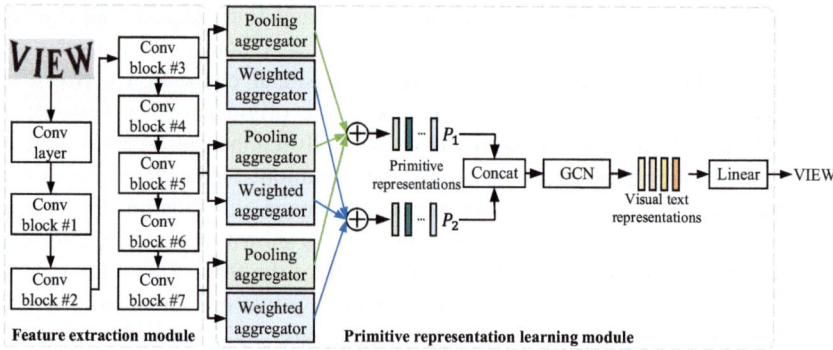

Fig. 2.6 Network structure of a primitive representation learning network with multiscale features

Fig. 2.6, the feature extraction module adopts EfficientNet-B3, which contains seven convolutional blocks. Assume that the number of channels of feature maps output by the final convolutional layer is d and that the number of primitive representations learned by each aggregator is n; then, each primitive representation has a size of $n \times d$.

According to the aggregating type, primitive representations generated by pooling aggregators and primitive representations generated by weighted aggregators are summed separately. The summation of primitive representations contains multiscale spatial information and retains the feature dimension d. Let the summation of primitive representations generated by pooling aggregators be $P_1 \in \mathbb{R}^{n \times d}$ and the summation of primitive representations generated by weighted aggregators be $P_2 \in \mathbb{R}^{n \times d}$, PREN concatenates P_1 and P_2 in the spatial dimension and generates the final primitive representations $P \in \mathbb{R}^{2n \times d}$.

The primitive representations P are transformed into visual text representations by a GCN, and a fully connected layer with the Softmax function is adopted to convert the visual text representations to probabilities for each character class. In this way, the PREN can decode all the characters in parallel.

2.4.3 Training and Inference

The PREN can be trained end-to-end. In the training stage, the cross-entropy between the predicted text and the ground-truth text is used as the objective function. The ground-truth text is obtained via the following two steps. First, an ⟨eos⟩ symbol is added after the last character of the label text to mark the end of decoding. Second, the padding symbol ⟨pad⟩ is used to extend the length of the text to the maximum

2.4 Primitive Representation Learning Network

decoding length L. Let l be the length of the original label text, the calculation formula of the objective function is

$$\mathcal{L} = -\frac{1}{l+1} \sum_{t=1}^{l+1} \log p(y_t | I) \tag{2.8}$$

where I is the input image, y_t ($t = 1, 2, \ldots, l$) is the t character of the label text, and y_{l+1} is the ⟨eos⟩ symbol. The padding symbol ⟨pad⟩ is not included in the calculation of the objective function. In the test stage, the PREN predicts L characters in parallel and takes the text before the first ⟨eos⟩ symbol as the final recognition result.

Moreover, the PREN has the ability to recognize text images with different orientation arrangements. While most English scene texts are horizontally arranged, some languages have many vertical texts, such as couplets in Chinese scenes. The diversity of text orientations introduces additional challenges for recognition.

Traditional sequence modeling methods convert input images into 1D feature sequences during feature extraction. Obviously, for vertical text images, it is not wise to compress the height of the feature map, so traditional methods usually rotate vertical input images by 90° and then recognize them as horizontal samples [5], as shown in Fig. 2.7a. However, the rotation of the input image causes the rotation of each character in the image, which doubles the number of patterns the network needs to learn. In addition, some characters after rotation have similar appearances to those of other characters, which introduces additional noise into the recognition process. In contrast, the PREN employs global feature aggregation to transform feature map output by the CNN into primitive representations, and this process is adapted to variable feature map sizes. As shown in Fig. 2.7b, input images with different text arrangements can maintain their original orientations, which increases the generalizability of the PREN to multioriented text images.

In the training stage, all the samples are divided into two subsets according to their aspect ratios. The samples in the horizontal subset and vertical subset have

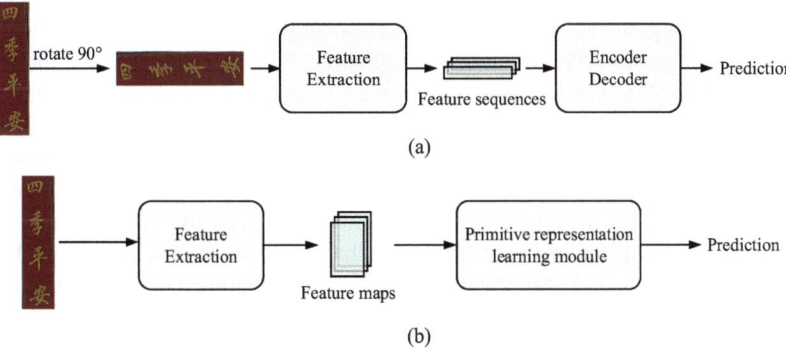

Fig. 2.7 The process of recognizing vertical text images by different models. (**a**) Traditional sequence-to-sequence methods. (**b**) PREN

different normalized sizes. The data of each training iteration are randomly taken from one of the subsets. The probability of selecting each subset equals the ratio of the number of remaining samples in one subset to the total number of remaining samples, i.e.,

$$p_H = \frac{n_H}{n_H + n_V} \tag{2.9}$$

$$p_V = \frac{n_V}{n_H + n_V} \tag{2.10}$$

where p_H and p_V are the probabilities of sampling from the horizontal and vertical subsets, respectively. n_H and n_V are the numbers of currently remaining samples in the horizontal and vertical subsets, respectively.

In the test stage, the horizontal and vertical orientations of the input image can be determined by setting a threshold according to the aspect ratio of the input image. In addition, the input image can be recognized with both horizontal normalization and vertical normalization, and the final prediction is the one with the highest recognition confidence.

2.5 Semantic-Guided Decoding

PREN learns primitive representations via global aggregations of feature maps output by a CNN, and then uses a GCN to obtain visual text representations. In this process, visual information in images is extracted and exploited. However, the semantic information contained in text has not been explored. The utilization of semantic information can further improve model performance. For example, attention-based encoder-decoder models often achieve higher accuracy than CTC-based models do. One of the reasons is that attention-based encoder-decoder models can learn an implicit language model during the decoding process [25].

To make better use of semantic information, this chapter proposes a semantic-guided decoding method that incorporates a language model and an attention module with the basic PREN model, as shown in Fig. 2.8. The language model is a Transformer [41] encoder and converts the visual text representations generated by the PREN into semantic representations. The attention module uses the semantic representations output by the language model to reweight the primitive representations to obtain visual-semantic representations, where the semantic representations are used as queries of the attention mechanism and the primitive representations are used as key-value pairs of the attention mechanism. Finally, the model combines three different types of features and generates semantic-guided decoding results.

2.5 Semantic-Guided Decoding

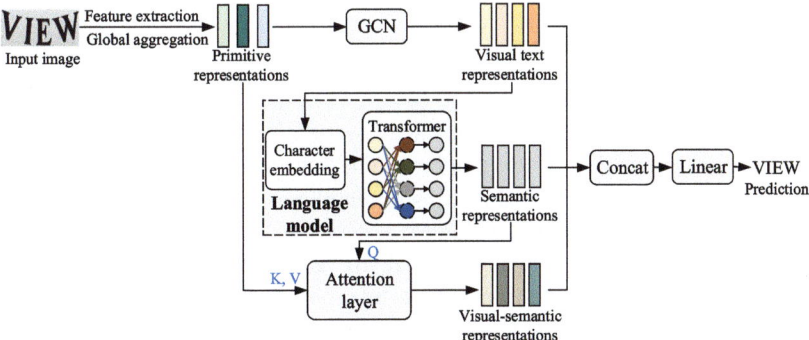

Fig. 2.8 Semantic-guided decoding method for the primitive representation learning network

2.5.1 Language Model with Attention Mask

The language model consists of a character embedding module and a Transformer encoder. First, visual text representations are transformed into character probabilities and then further mapped to a set of character embedding vectors through fully connected layers. Let $V \in \mathbb{R}^{L \times d}$ be the visual text representation learned by the PREN; then, the calculation process of character embedding is

$$B = \text{softmax}(V W_v) \tag{2.11}$$

$$E = B W_e + \boldsymbol{b}_e \tag{2.12}$$

where $B \in \mathbb{R}^{L \times C}$ denotes the character probabilities. $E \in \mathbb{R}^{L \times d}$ denotes the character embedding vectors. Using $W_v \in \mathbb{R}^{d \times C}$ and the Softmax function converts visual text representations into character probabilities, where C is the number of classes. $W_e \in \mathbb{R}^{C \times d}$ and $\boldsymbol{b}_e \in \mathbb{R}^d$ transform the character probabilities into character embedding vectors.

In general, language models can be pretrained on a text corpus. One simple way is to use one-hot encodings of ground-truth text to replace the character probabilities B. However, as shown in Eq. (2.12), the values of the character probabilities B are computed from visual text representations and have different distributions with one-hot encodings. To solve this problem, this chapter proposes modeling the character probabilities predicted by the visual text representation as a multivariate Gaussian distribution. For a pretrained PREN model, the predicted character probabilities for all samples can be calculated first, and then the mean and covariance of each character class can be calculated, as shown in Eqs. (2.13) and (2.14).

$$\boldsymbol{\mu}_c = \frac{1}{N_c} \sum_{i=1}^{N_c} \boldsymbol{p}_i^{(c)} \tag{2.13}$$

$$\Sigma_c = \frac{1}{N_c - 1} \sum_{i=1}^{N_c} \left(\boldsymbol{p}_i^{(c)} - \boldsymbol{\mu}_c\right)^T \left(\boldsymbol{p}_i^{(c)} - \boldsymbol{\mu}_c\right) \tag{2.14}$$

where $\boldsymbol{\mu}_c$ and Σ_c are the mean vector and covariance matrix corresponding to the c-th class, respectively. N_c is the number of characters that are predicted as the c-th class. $\boldsymbol{p}_i^{(c)} \in \mathbb{R}^{1 \times C}$ is the probability distribution of a character that is predicted as the c-th class. C is the number of classes.

When pretraining the language model, for each character in the input text, assuming that its class is c, the input character probabilities are obtained via random sampling from a multivariate Gaussian distribution with mean $\boldsymbol{\mu}_c$ and covariance Σ_c.

In the language model, the positional information is injected into the character embedding vectors. A commonly used positional encoding strategy is the sinusoidal position encoding proposed by Vaswani et al. [41], which can be formulated as

$$PE_{(t,2i)} = \sin\left(t/10{,}000^{2i/d}\right) \tag{2.15}$$

$$PE_{(t,2i+1)} = \cos\left(t/10{,}000^{2i/d}\right) \tag{2.16}$$

where t refers to the position of each character and where $2i$ and $2i+1$ are the indices of the feature dimension. Finally, the model adds character embedding vectors and positional encodings as the inputs of the Transformer encoder, i.e., $\widetilde{E} = E + PE$.

The Transformer encoder in the language model predicts the probability of a character through its context, with the goal of

$$\max \frac{1}{|\mathcal{T}|} \sum_{t \in \mathcal{T}} p\left(y_t | y_1, \ldots, y_{t-1}, y_{t+1}, \ldots, y_{|\mathcal{T}|}\right) \tag{2.17}$$

where $\mathcal{T} = \{1, 2, \ldots, l+1\}$ is the set of character positions to be predicted, y_t is the t character in the ground-truth text, and y_{l+1} is the ⟨eos⟩ symbol.

Language models based on Transformer encoders can generally be divided into two main types: autoencoding language models and autoregressive language models. Autoencoding language models, such as BERT [7] and RoBERTa [24], replace some of the input tokens with masks and predict the characters or words in the masked positions. On the other hand, autoregressive language models, such as XLNet [45], do not add masks to input tokens directly. Instead, some of the attention weights are masked during the computation process of the self-attention mechanism. In this way, the model can avoid using a token when predicting itself. This chapter adopts the autoregressive language model because it does not introduce the extra mask symbol in the alphabet. Because the number of characters in scene text images is usually limited, the model masks only the attention weight corresponding to each

2.5 Semantic-Guided Decoding

character to be predicted. In this way, the features of the character to be predicted can be updated with the features of all the remaining characters.

Specifically, the Transformer encoder is composed of several identical blocks. Each block uses the output of the previous block as a query and uses the output of the character embedding module $\widetilde{E}$ as key-value pairs. The calculation process of the Transformer encoder is as follows:

$$O_l = \text{softmax}\left(\frac{1}{\sqrt{d}} O_{l-1} W_Q^{(l)} W_K^{(l)T} \widetilde{E}^T + M\right) \widetilde{E} W_V^{(l)} \quad (2.18)$$

where O_l represents the output of the l-th block, $W_Q^{(l)}$, $W_K^{(l)}$ and $W_V^{(l)}$ are learnable mapping matrices in the l-th block. For the first block, the input O_0 consists of learnable vectors that are randomly initialized and trained with the network. $M = \{m_{ij}\}_{i,j=0}^{L}$ denotes attention masks, which are composed of $m_{ii} = -\infty$ ($\forall i = 1, 2, \ldots, L$), and $m_{ij} = 0$ ($\forall j \neq i$).

In this way, the weights of all the remaining characters are used to update the features of the current character. The output of the final block in the Transformer encoder is taken as the semantic representation S.

2.5.2 Fusion of Visual and Semantic Information

Fang et al. [8] proposed a text image recognition method based on parallel attention mechanism and language modeling, which also uses a Transformer encoder with a diagonal attention mask to improve the performance of the attention-based encoder-decoder model. In contrast, the semantic-guided decoding method proposed in this chapter uses semantic representations output by the language model to further reweight primitive representations learned by the PREN. In the original PREN model, primitive representations are weighted via combination coefficients (the adjacency matrix of the GCN), which are parameters trained with the network and are fixed after the training stage. Therefore, the combination coefficients of primitive representations in the GCN are the same for different input images. Instead, the semantic-guided decoding method allows the primitive representations to have different combinations for different input images, which makes the decoding more flexible.

The above process can be implemented by an attention module. The semantic representations S are used as queries, and the primitive representations are used as key-value pairs, i.e.,

$$G = \text{softmax}\left(\frac{1}{\sqrt{d}} S W_Q W_K^T P^T\right) P W_V \quad (2.19)$$

where $P \in \mathbb{R}^{n \times d}$ denotes primitive representations and where W_Q, W_K and W_V are learnable mapping matrices. Since semantic representations S contain rich semantic information and primitive representations P contain the global visual information of the image, the output G of the attention module combines both visual and semantic information. Therefore, G is called a visual-semantic representation.

The visual text representations, semantic representations, and visual-semantic representations can be used together to obtain the final decoding results. A fully connected layer is used to integrate the three representations, as shown in Eq. (2.20).

$$\mathcal{P} = \text{softmax}\left(\phi\left([V; S; G]W_r\right)W_o\right) \tag{2.20}$$

where V, S, and G are visual text representations, semantic representations and visual semantic representations, respectively; $[V; S; G]$ denotes the concatenation operation; and $W_r \in \mathbb{R}^{3d \times d}$ and $\phi(\cdot)$ are the mapping matrix and activation function of the fully connected layer. Finally, a linear transformation $W_o \in \mathbb{R}^{d \times C}$ and the Softmax function are used to convert integrated representations into character probabilities $\mathcal{P}$. C is the number of classes.

In the training stage, the objective function of the model is

$$\mathcal{L} = \mathcal{L}_f + \lambda_1 \mathcal{L}_v + \lambda_2 \mathcal{L}_s + \lambda_3 \mathcal{L}_g \tag{2.21}$$

where $\mathcal{L}_v$, $\mathcal{L}_s$, $\mathcal{L}_g$ and $\mathcal{L}_f$ are the cross-entropy losses of the ground-truth text and character probabilities calculated by the visual text representations, the semantic representations, the visual-semantic fusion representations, and the integrated representations of the three, respectively. $\{\lambda_i\}_{i=1}^{3}$ are used to balance different objective functions. In the experiments, λ_1, λ_2 and λ_3 are all set to 1. In the test stage, the decoded results computed by the integrated representations are used as the final prediction.

2.6 Discussion

This section discusses the effects of different modules in the PREN and compares the PREN with other widely used text recognition models on datasets of different languages.

2.6.1 English Scene Text Recognition

2.6.1.1 Experimental Settings

For English scene text recognition, all the models are trained using the MJSynth [16] and SynthText [10] synthetic scene text image datasets. The test set includes seven public real scene text image datasets, i.e., IIIT5K [28], SVT [42], IC03 [27], IC13 [17], IC15 [18], SVTP [30] and CUTE [33].

2.6 Discussion

All the models are trained for five epochs, the initial learning rate is set to 5×10^{-4}, and the learning rate decreases to 0.5, 0.1, and 0.05 in the second, fourth, and fifth epochs, respectively. The batch size is set to 512. AdamW [26] is adopted as the optimizer. The input image is normalized to 64×256 pixels. The alphabet to be recognized includes 26 case-insensitive English letters and 10 numbers. Since the maximum length of words in the dataset used in this book does not exceed 25, the maximum decoding length is set to 25 in the experiments. The evaluation index is word accuracy.

2.6.1.2 Number of Primitive Representations

This experiment compares the performance of models that learn different numbers of primitive representations. All the models use ResNet-50 for the feature extraction module, and only the feature maps of the final convolutional layer are used for learning primitive representations.

Table 2.1 lists the word accuracies of the models as the number of primitive representations increases from 1 to 10 on seven English scene text datasets. In Table 2.1, "#Primitive representations" refers to the number of primitive representations learned by each global aggregator in the PREN model. Since PREN concatenates the primitive representations learned by pooling aggregators and weighted aggregators, the actual number of primitive representations is twice the quantity listed in the table. "Average" refers to the overall average word accuracy of the model over the seven test sets, which is calculated as the ratio of the number of all correctly recognized images to the number of all test samples. As the number of primitive representations increases, the performance of the PREN first increases but then decreases. When the number of primitive representations is 5, the model achieves the highest average word accuracy on the test sets. According to the results

Table 2.1 Word accuracy (%) of models with different numbers of primitive representations

#Primitive representations	IIIT5K	SVT	IC03	IC13	IC15	SVTP	CUTE	Average
1	90.6	86.4	92.0	92.3	79.0	79.7	80.9	86.8
2	90.9	88.6	92.8	92.8	79.1	79.8	83.0	87.3
3	91.0	89.3	**93.9**	93.8	79.2	79.2	84.4	87.7
4	91.2	88.6	93.5	93.5	79.6	79.2	**85.1**	87.7
5	91.7	88.9	93.3	**94.2**	79.7	**82.5**	84.7	**88.3**
6	**92.0**	**90.1**	93.0	93.5	79.8	80.9	83.7	88.2
7	91.1	**90.1**	93.5	**94.2**	80.1	80.5	84.7	88.1
8	90.9	89.6	93.7	93.8	**80.3**	81.1	**85.1**	88.0
9	90.9	89.3	**93.9**	**94.2**	80.2	80.2	84.7	88.0
10	90.8	**90.1**	93.0	93.5	80.1	80.2	83.7	87.8

Bold values highlight the optimal performance (e.g., highest accuracy or lowest error rate) for the respective metric across different methods, models, or systems

in Table 2.1, the number of primitive representations learned by each aggregator is set to 5 in the subsequent experiments and discussions.

2.6.1.3 Comparison of Feature Aggregation Methods

This experiment compares the performance of models using different feature aggregation methods. All the models use ResNet-50 as the feature extraction module, and only the feature maps of the final convolutional layer are used for learning primitive representations.

Table 2.2 lists the word accuracy of models with different feature aggregation methods, including the use of pooling aggregators or weighted aggregators individually and jointly. In Table 2.2, "Pooling + Weighting" means that primitive representations learned by pooling aggregators and weighted aggregators are summed to obtain the final primitive representations, and "Pooling —— Weighting" means that primitive representations learned by pooling aggregators and weighted aggregators are concatenated in the spatial dimension. Table 2.2 shows that the combination of pooling aggregators and weighted aggregators can improve model performance. The concatenation of primitive representations learned by pooling aggregators and weighted aggregators achieves the highest word accuracy.

Table 2.2 further shows that, compared with the model with only weighted aggregators, the model with only pooling aggregators achieves better performance on regular text image datasets (IIIT5K, SVT, IC03 and IC13) but has lower performance on irregular text image datasets (IC15, SVTP and CUTE). The results imply that the pooling aggregator is more conducive to the recognition of regular text images by sharing the aggregating weights for different samples, whereas the spatial attention mechanism adopted in the weighted aggregator is more suitable for recognizing irregular text images with redundant background information.

Figure 2.9 shows the average word accuracy on seven test sets of models with different feature aggregation methods and numbers of primitive representations. The results indicate that models that use both pooling aggregators and weighted aggregators outperform models that use only a single type of feature aggregator. The "Pooling —— Weighting" model achieves the best performance (88.3%) when each aggregator learns five primitive representations, whereas the

Table 2.2 Word accuracy (%) of models with different feature aggregation methods

Aggregation	IIIT5K	SVT	IC03	IC13	IC15	SVTP	CUTE	Average
Pooling	91.0	88.1	92.5	93.0	78.7	78.9	83.3	87.2
Weighted	90.9	87.5	92.4	93.0	79.6	79.5	84.4	87.4
Pooling + Weighted	90.7	**89.5**	93.1	93.2	79.3	81.9	**85.1**	87.7
Pooling —— Weighted	**91.7**	88.9	**93.3**	**94.2**	**79.7**	**82.5**	84.7	**88.3**

Bold values highlight the optimal performance (e.g., highest accuracy or lowest error rate) for the respective metric across different methods, models, or systems

2.6 Discussion

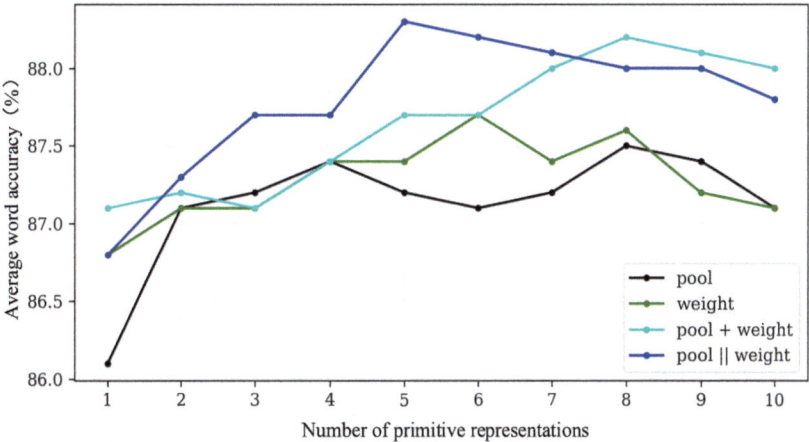

Fig. 2.9 Effects of the feature aggregation method and the number of primitive representations on model performance

"Pooling + Weighting" model achieves its best performance (88.2%) when each aggregator learns eight primitive representations. It can be inferred that concatenation can better preserve visual information in primitive representations.

2.6.1.4 Comparison of Feature Extraction Methods

This experiment compares the performance of models with different CNN backbone networks (ResNet-50 and EfficientNet-B3), and explores the effect of using multi-scale feature maps on the recognition accuracy. Table 2.3 lists the word accuracy of the PREN with different feature extraction methods on seven test sets. In Table 2.3, F_i denotes the feature map output by the i-th convolution block. For example, F_{234} refers to using outputs of the second, third and fourth convolution blocks in ResNet-50, and F_{357} refers to using outputs of the third, fifth and seventh convolutional blocks in EfficientNet-B3. F_2, F_3, and F_4 of ResNet-50 and F_3, F_5, and F_7 of EfficientNet-B3 are feature maps with spatial sizes of $\frac{1}{8}$, $\frac{1}{16}$ and $\frac{1}{32}$ of the input image, respectively.

The following phenomena can be observed from Table 2.3.

1. The feature maps output by different convolutional blocks of a CNN significantly influence the recognition accuracy. When features from shallow convolutional layers (e.g., F_2 of ResNet-50 or F_3 of EfficientNet-B3) are used, the word accuracy of the model is relatively low. When features from deep convolutional layers (e.g., F_4 for ResNet-50 or F_7 for EfficientNet-B3) are used, the model achieves higher word accuracy. Using F_4 of ResNet-50 for primitive representation learning achieves an average word accuracy improvement of 9.6% over using F_2.

Table 2.3 Word accuracy (%) of models with different feature extraction methods

CNN	Feature maps	IIIT5K	SVT	IC03	IC13	IC15	SVTP	CUTE	Average
ResNet-50	F_2	83.1	79.8	87.5	86.7	68.5	67.3	68.4	78.7
	F_3	90.3	87.2	91.2	91.7	77.7	78.5	80.6	86.2
	F_4	91.7	88.9	93.3	94.2	79.7	82.5	84.7	88.3
	F_{23}	91.3	87.0	92.3	92.2	78.4	78.9	83.7	87.0
	F_{24}	91.2	90.1	93.8	93.6	80.0	80.8	84.0	88.1
	F_{34}	91.9	89.6	93.1	94.5	80.2	81.7	83.7	88.2
	F_{234}	92.2	90.6	93.8	94.6	80.7	82.8	**86.1**	89.0
EfficientNet-B3	F_3	77.0	73.4	84.2	82.6	64.3	62.8	51.0	73.2
	F_5	91.8	88.9	93.5	93.7	82.1	81.2	76.0	88.4
	F_7	92.4	91.8	95.0	95.9	83.1	**83.4**	78.5	89.7
	F_{35}	91.5	89.6	93.1	94.0	81.8	80.8	79.9	88.4
	F_{37}	92.2	92.4	94.1	94.9	**83.9**	**83.4**	80.9	89.8
	F_{57}	92.1	**92.7**	94.9	**96.1**	**83.9**	83.3	80.1	**89.9**
	F_{357}	**92.5**	92.3	**95.2**	**96.1**	83.2	**83.4**	80.2	**89.9**

Bold values highlight the optimal performance (e.g., highest accuracy or lowest error rate) for the respective metric across different methods, models, or systems

2. Using multiscale feature maps can improve model performance. For both ResNet-50 and EfficientNet-B3, when three feature maps with different scales are used, the average word accuracy of the model can be improved by 0.7% and 0.2%, respectively, compared with models that use only feature maps from the final convolutional layer.
3. Using EfficientNet-B3 as the feature extraction module achieves higher recognition accuracy than does using ResNet-50 on most test sets, which is in accordance with the result that EfficientNet-B3 has better performance than ResNet-50 on general object recognition tasks.

2.6.1.5 Comparison of Different Models

This experiment compares three different scene text recognition frameworks: CTC-based methods, attention-based encoder-decoder methods, and the PREN. For the CTC-based methods, the CRNN model [35] is adopted. For the attention-based encoder-decoder methods, the ASTER model [37] is adopted. For fair comparison, the three models use the same CNN as the feature extraction module, i.e., EfficientNet-B3. The CRNN and ASTER models with EfficientNet-B3 are denoted as CNN-LSTM-CTC and Attention-1D, respectively. CNN-LSTM-CTC, Attention-1D, and the PREN share the same training configurations.

Table 2.4 lists the word accuracy of each model on seven test sets and the average time it takes to recognize an image. In terms of word accuracy, the PREN achieves better performance on all test sets than does the CNN-LSTM-CTC. In particular, the accuracy of the PREN on the SVTP irregular dataset exceeds that of the CNN-

2.6 Discussion

Table 2.4 Word accuracy (%) and average recognition time of different models

Model	IIIT5K	SVT	IC03	IC13	IC15	SVTP	CUTE	Time
CRNN [35]	82.9	81.6	93.1	89.2	64.2	70.0	65.5	–
ASTER [37]	93.4	89.5	94.5	91.8	76.1	78.5	79.5	–
CNN-LSTM-CTC	92.0	88.7	92.8	93.0	79.9	78.3	78.5	23.6 ms
Attention-1D	**94.2**	91.5	94.7	96.0	**84.4**	**83.4**	**83.7**	33.9 ms
PREN	92.5	**92.3**	**95.2**	**96.1**	83.2	**83.4**	80.2	22.7 ms

Bold values highlight the optimal performance (e.g., highest accuracy or lowest error rate) for the respective metric across different methods, models, or systems

LSTM-CTC model by 5.1%, indicating that the primitive representation learning method can learn richer 2D visual information. The Attention-1D model achieves the best performance on the three test sets (IIIT5K, IC15, and CUTE), indicating that the implicit language model learned by the attention mechanism is important for recognizing some difficult samples.

For recognition speed, CTC-based methods need to ensure that the length of the feature sequence output by the encoder cannot be less than the number of characters to be recognized; otherwise, the objective function cannot be calculated in the training stage because there is no decoding path that can generate the ground-truth text. In the test stage, the model also fails to predict the correct results. Attention-based encoder-decoder methods follow a recursive decoding process, making the average recognition time of Attention-1D larger than that of CNN-LSTM-CTC and PREN. Instead, each feature aggregator in the PREN needs to learn only five primitive representations, and additional computational complexity, such as upsampling, convolution or pooling, is not introduced when multiscale feature maps are used. Therefore, the PREN has the fastest recognition speed.

The results in Table 2.4 demonstrate that, compared with the commonly used CTC-based methods and attention-based encoder-decoder methods, the PREN has both higher recognition accuracy and recognition efficiency.

2.6.1.6 Visualization

To explore the information contained in the primitive representations, this chapter visualizes some intermediate features generated by feature aggregators. For the pooling aggregator, Fig. 2.10a shows feature maps before pooling (i.e., after $conv_2(\cdot)$ and before $Pool(\cdot)$ in Eq. (2.3)). Since primitive representation is obtained by global average pooling of the feature maps, the position with a higher value in the feature maps contributes more to the primitive representation, indicating that more attention is given to this region in the primitive representation. As shown in Fig. 2.10a, feature maps corresponding to the same primitive representation are similar; e.g., for the two input images, the bottom left parts of the first feature maps have higher values.

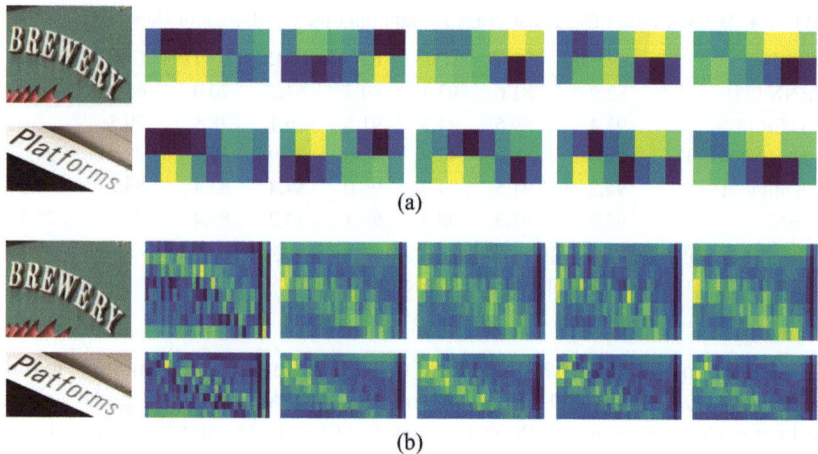

Fig. 2.10 Visualization of different aggregation methods in scene text recognition tasks. (**a**) Visualization of feature maps before average pooling in the pooling aggregator. (**b**) Visualization of aggregating weights in the weighted aggregator

Figure 2.10b shows heatmaps generated in the weighted aggregator, which are used as aggregating weights. Thus, positions in feature maps with larger aggregating weights contribute more to the primitive representation. As shown in Fig. 2.10b, aggregating weights typically exhibit larger values in the foreground regions of the images and smaller values in the background areas. This finding indicates that weighted aggregators can reduce the influence of redundant background noise on the recognition process.

The above analysis may explain why the model with only pooling aggregators achieves higher word accuracy on regular scene text datasets, whereas the model with only weighted aggregators achieves higher word accuracy on irregular scene text datasets. Since the two aggregation methods are complementary, integrating two types of aggregators can improve model performance, as shown in Table 2.2.

Table 2.5 lists some recognition examples of different models, and the recognition errors are marked in red. Samples with complex backgrounds and text layouts are challenging for CNN-LSTM-CTC and Attention-1D. In contrast, the PREN can better address background noise, ligatures and irregular text layouts. However, as shown in the fourth sample in Table 2.5, since the PREN uses only visual information for recognition, low-quality images easily cause recognition errors. Attention-based encoder-decoder methods can alleviate this problem by learning an implicit language model in the decoding process.

Figure 2.11 shows more failure cases of the PREN in the English scene text recognition task. Two typical types of errors include text images with special fonts and low-quality images such as motion blur.

2.6 Discussion

Table 2.5 Recognition examples of different models on English scene text images

Input image				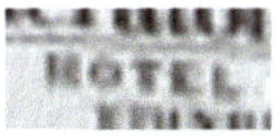
Difficulty	Background noise	Ligatures	Irregular layout	Blur
Ground-truth text	beijing	hortons	hadalabo	hotel
CNN-LSTM-CTC	beljing	hoilons	_adalabo	_otre_
Attention-1D	belling	hotloads	_adalabo	hotel
PREN	beijing	hortons	hadalabo	horee

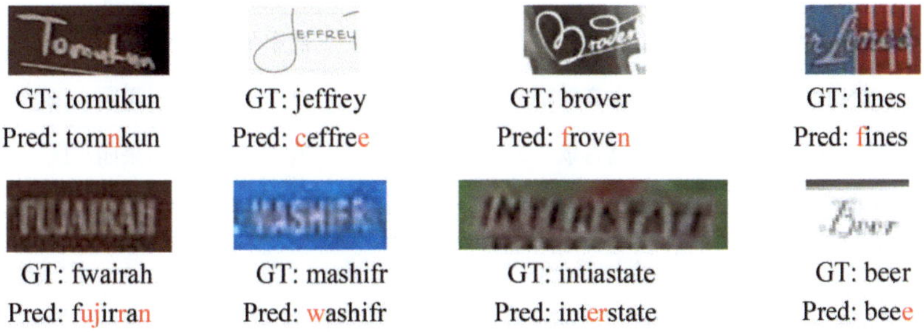

GT: tomukun
Pred: tomnkun

GT: jeffrey
Pred: ceffree

GT: brover
Pred: froven

GT: lines
Pred: fines

GT: fwairah
Pred: fujirran

GT: mashifr
Pred: washifr

GT: intiastate
Pred: interstate

GT: beer
Pred: beee

Fig. 2.11 Failure cases of the primitive representation learning network in the English scene text recognition task

2.6.2 Multioriented Chinese Scene Text Recognition

Compared with English scene text images, Chinese images have thousands of commonly used characters, including many similar characters. In addition, multi-oriented texts (i.e., both horizontal and vertical texts) are common in Chinese scene images, which introduces additional challenges to the recognition task. Therefore, Chinese scene text recognition is a challenging task in which the robustness of scene text recognition can be evaluated.

All the models are first trained on the SynthChinese synthetic Chinese scene text dataset for 6 epochs and then fine-tuned for 20 epochs on a subset of the RCTW dataset [36]. The learning rate is initialized to 0.5 and reduced to 0.1 in the sixth epoch, and the optimizer is ADADELTA [47]. All the models use EfficientNet-B3 [39] as the feature extraction module.

In the training stage, for the CNN-LSTM-CTC model and the Attention-1D model, vertical text images are rotated 90° as the input of the network, and all input images are normalized to 64 × 256 pixels. For PREN, horizontal samples are normalized to 64 × 256 pixels, and vertical samples are normalized to 256 × 64 pixels.

Table 2.6 Word accuracy (%) of different models in the multioriented Chinese scene text recognition task

Model	Horizontal	Vertical	Average
CNN-LSTM-CTC	59.8	67.8	63.8
Attention-1D	68.2	75.0	71.6
PREN	**74.0**	**79.2**	**76.6**

Bold values highlight the optimal performance (e.g., highest accuracy or lowest error rate) for the respective metric across different methods, models, or systems

Table 2.7 Recognition examples of different models on Chinese scene text images

Input image				
Ground-truth text	女高音歌唱家	来自讯道	四季平安	密云果园小区
CNN-LSTM-CTC	女高音歌唱务	来自保道	四季平茂	密云果园小区
Attention-1D	女高音喝家	来自沃德	四季平安	恋云果园小区
PREN	女高音歌唱家	来自讯道	四季平安	密云果园小区

Table 2.6 lists the performance of different models on the multioriented Chinese scene text recognition task. The PREN significantly outperforms the CNN-LSTM-CTC and Attention-1D models. For the CNN-LSTM-CTC and Attention-1D models, rotation of vertical text images introduces extra noise into the recognition process, and the PREN can avoid this problem. The experimental results demonstrate the superiority of the PREN in recognizing multioriented scene text images (Table 2.7).

To gain a deeper understanding of primitive representations, an additional Chinese character recognition experiment is conducted on the historical Chinese character dataset [3]. Figure 2.12 shows the visualization results of different aggregation methods. For the pooling aggregator, Fig. 2.12a shows similar results as Fig. 2.10a, i.e. for different input images, primitive representations learned by pooling aggregators focus on similar regions. For the weighted aggregator, different primitive representations focus on specific strokes in characters. As shown in Fig. 2.12b, the first and third primitive representations focus on horizontal strokes in the input images, whereas the second primitive representation focuses on vertical strokes in the input images. The fifth primitive representation focuses on the whole foreground area of the input images.

According to the above visualizations, pooling aggregation learns the intrinsic information from different text instances. Weighted aggregation can effectively distinguish image foreground and background regions and learn specific stroke information in characters.

2.6 Discussion

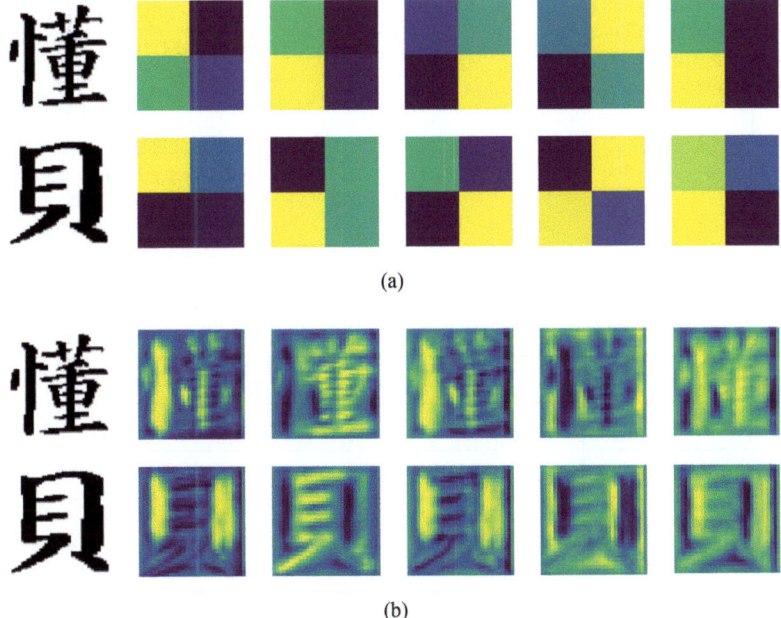

(a)

(b)

Fig. 2.12 Visualization of different aggregation methods in the Chinese character recognition task. (**a**) Visualization of feature maps before average pooling in the pooling aggregator. (**b**) Visualization of aggregating weights in the weighted aggregator

2.6.3 Arabic Handwriting Recognition

In addition to the scene text recognition task, this chapter also verifies the effectiveness of the PREN on the handwritten Arabic word recognition task. There are many cursive scripts and ligatures in handwritten Arabic. Additionally, there are numerous similar characters.Therefore, handwritten Arabic recognition tasks require models to have strong spatial information modeling capabilities.

All the models are trained on subsets a, b, c, and d of the IFN/ENIT handwritten Arabic dataset [29] for 100 epochs, and tested on subset e. The learning rate is initialized to 1×10^{-3} and decreases 0.5 times for every 10 training epochs from the 25th epoch. The optimizer is AdamW [26]. All the input images are normalized to 64×256 pixels. Because the reading order of Arabic is from right to left, the label text is reversed during model training so that the models decode from left to right. The decoding result is subsequently reversed again to obtain the final prediction. All the models adopt EfficientNet-B3 as the feature extraction module. The character set contains 120 characters, including the Arabic alphabet and numbers. The evaluation metrics include the word error rate (WER) and the character error rate (CER).

Table 2.8 lists the comparison of different models in the handwritten Arabic recognition task. For both WER and CER, the PREN can achieve significantly

Table 2.8 Performance (%) of different models in the handwritten Arabic recognition task

Model	CER	WER
CNN-LSTM-CTC	6.61	27.8
Attention-1D	10.64	23.3
PREN	**4.58**	**17.5**

Bold values highlight the optimal performance (e.g., highest accuracy or lowest error rate) for the respective metric across different methods, models, or systems

Table 2.9 Examples of recognition results of handwritten Arabic text images by different models

Input image	غزالة تمرة دار علّوش المنار2
Ground-truth text	غزالة تمرة دارعلوش المنار2
CNN-LSTM-CTC	غرانة تمزة دارعلوش الم ض ار2
Attention-1D	زالة حزوة بو ع ط وش المنار2
PREN	غزالة تمرة دارعلوش المنار2

lower results, indicating that primitive representation learning can better handle handwritten words than traditional sequence modeling methods can. Compared with CNN-LSTM-CTC, Attention-1D with attention mechanism achieves a lower WER because the implicit language model learned by the attention mechanism is beneficial for recognizing similar characters, and there are more correctly recognized words. However, Attention-1D has a higher CER because the ligatures in handwritten text images make it difficult for the attention-based decoder to accurately align each character with features, resulting in the attention drift problem. Compared with that of Attention-1D, the WER of the PREN is reduced by 5.8%; compared with that of CNN-LSTM-CTC, the CER of the PREN is reduced by 2.03%.

Table 2.9 shows recognition examples of different models on handwritten Arabic images. CNN-LSTM-CTC easily confuses similar characters. Attention-1D can distinguish similar characters better than can CNN-LSTM-CTC but sometimes produces more serious recognition errors, as shown in the third sample in Table 2.9. Compared with CNN-LSTM-CTC and Attention-1D, the PREN has better recognition ability for difficult situations such as similar characters and ligatures.

2.6.4 Semantic-Guided Decoding Experiment

This experiment explores the effect of the semantic-guided decoding method. The PREN and the language model are first pretrained separately and then combined.

2.6 Discussion

Table 2.10 Word accuracy (%) of different models

Model	IIIT5K	SVT	IC03	IC13	IC15	SVTP	CUTE	Average
PREN	91.7	88.9	93.3	94.2	79.7	82.5	84.7	88.3
PREN + Language modeling	93.2	92.7	94.7	96.0	82.3	**86.8**	86.5	90.4
PREN + Semantic-guided decoding	**93.4**	**92.9**	**94.9**	**96.4**	**82.9**	86.5	**88.5**	**90.8**

Bold values highlight the optimal performance (e.g., highest accuracy or lowest error rate) for the respective metric across different methods, models, or systems

PREN uses ResNet-50 as the feature extraction module, and only uses feature maps of the final convolutional layer for primitive representation learning. The training configurations are the same as those in Sect. 2.6.1. The language model is a four-layer Transformer encoder.

For language model pretraining, all ground-truth texts in the MJSynth [16] and SynthText [10] datasets are extracted as corpora, which contain 198,548 different words. When pretraining the language model, the batch size is set to 128, and the number of training epochs is set to 200. The learning rate is initialized to 1×10^{-4}, and reduced to 1×10^{-5} at the 80th epoch. The optimizer is AdamW [26].

When combining the PREN and the language model, the model is trained on the MJSynth [16] and SynthText [10] datasets for five epochs. The optimal learning rates for the PREN and the language model are different in preliminary experiments. Therefore, for the feature extraction module and the primitive representation learning module, the learning rate is initialized as 1×10^{-3}. For the language model, the learning rate is initialized as 1×10^{-4}. The learning rates of all the modules are reduced by factors of 0.5, 0.1, and 0.05 at the second, fourth, and fifth epochs, respectively.

Table 2.10 lists the word accuracies of the models before and after introducing semantic-guided decoding. In Table 2.10, "PREN" is the original model that uses only visual text representations for decoding. "PREN + Language modeling" is a model that combines visual text representations and semantic representations output by the language model but does not use semantic representations to reweight primitive representations. "PREN + Semantic-guided decoding" is a model that combines visual text representations, semantic representations and visual-semantic representations.

The performance of the PREN significantly improves when the language model is combined, and the average word accuracy increases from 88.3% to 90.4%. After semantic representations are used to reweight primitive representations, the word accuracy of the model on most test sets is further improved, and the average word accuracy is increased to 90.8%. The experimental results demonstrate the effectiveness of the semantic-guided decoding method.

PREN with semantic-guided decoding generates visual text representations, semantic representations and visual-semantic representations in the calculation pro-

Table 2.11 Word accuracy (%) when decoding with different representations

Representation	IIIT5K	SVT	IC03	IC13	IC15	SVTP	CUTE	Average
Visual text representations	92.3	90.3	93.5	94.3	80.6	82.3	86.5	88.9
Semantic representations	43.8	59.2	60.7	64.5	36.9	54.0	38.9	48.1
Visual-semantic representations	91.2	**93.8**	94.6	95.4	81.4	**87.0**	86.5	89.5
Integrated representations	**93.4**	92.9	**94.9**	**96.4**	**82.9**	86.5	**88.5**	**90.8**

Bold values highlight the optimal performance (e.g., highest accuracy or lowest error rate) for the respective metric across different methods, models, or systems

cess. To explore the effects of different representations, this experiment compares the word accuracy of the same model using different representations for decoding, as shown in Table 2.11. "Integrated representations" means using the integration of all three types of representations for decoding. The following phenomena can be observed:

1. After introducing the semantic-guided decoding method, the performance of PREN itself can be improved slightly. Compared with the results in Table 2.10, PREN with semantic-guided decoding increases the average word accuracy from 88.3% to 88.9% when only visual text representations are used for decoding.
2. It is difficult to accurately recognize scene text images with only semantic representations because text images usually contain texts outside the corpus. The average word accuracy is 48.1% when only semantic representations are used for decoding. However, semantic representations can help improve the performance of visual text representations. Compared with using only visual text representations, the average word accuracy improved by 0.6% when visual-semantic representations were used for decoding.
3. The integration of three types of representations achieves the best results, and increases the average word accuracy to 90.8%.

Table 2.12 shows recognition examples of models that use different representations for decoding. In the first sample, decoding with only semantic representations results in incorrect predictions. When "ish" is used to predict the first character, "**f**ish" and "**w**ish" are both reasonable inferences since there is no more contextual information. Errors caused by semantic uncertainty lead to low word accuracy when decoding with only semantic representations. The second sample shows that the error of visual text representations is corrected by semantic representations. Since the input image is very low-quality, using visual text representations alone results in the incorrect prediction of "sensitvv". However, with semantic representations, the prediction is corrected such that it is more reasonably "sensitive". In the third sample, either using visual text representations or using semantic representations leads to errors. However, using visual-semantic representations for decoding yields

2.6 Discussion

Table 2.12 Recognition examples of models that use different representations for decoding

Input image	FISH	SENSITIVE	Center	SHAKESPEARE
Ground-truth text	fish	sensitive	center	shakespeare
Visual text representations	fish	sensit_vv	denter	smakespeare
Semantic representations	wish	sensitive	censed	_hokespeare
Visual-semantic representations	fish	sensitive	center	shokespeare
Combination of three representations	fish	sensitive	center	shakespeare

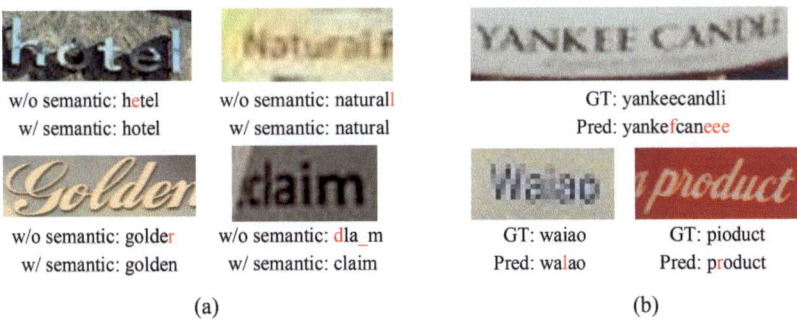

w/o semantic: hetel
w/ semantic: hotel

w/o semantic: naturall
w/ semantic: natural

GT: yankeecandli
Pred: yankefcaneee

w/o semantic: golder
w/ semantic: golden

w/o semantic: dla_m
w/ semantic: claim

GT: waiao
Pred: walao

GT: pioduct
Pred: product

(a)   (b)

Fig. 2.13 More recognition examples of semantic-guided decoding. (**a**) Comparison of recognition results before and after introducing semantic-guided decoding. (**b**) Failure cases of the PREN with semantic-guided decoding

correct results. The last sample shows that three representations are integrated to obtain the correct prediction. The results in Table 2.12 demonstrate that the semantic-guided decoding method can effectively improve the performance of the model.

Figure 2.13a shows some examples that the original PREN fails to recognize, but the PREN with semantic-guided decoding yields correct predictions. Semantic-guided decoding is helpful for recognizing special fonts and low-quality images and can also correct missing or redundant characters.

Figure 2.13b shows incorrectly recognized samples for the PREN with semantic-guided decoding. Errors easily occur when the input image lacks semantic information and is blurred. In addition, images with long texts are still challenging for the PREN.

2.7 Summary

This chapter proposes a primitive representation learning method. Unlike existing CTC-based models and attention-based encoder-decoder models, primitive representation learning converts feature maps extracted by a CNN into primitive representations through global feature aggregation. A graph convolutional network is further used to transform primitive representations into visual text representations, which can be used for parallel decoding. For global feature aggregation, a pooling aggregation method and a weighted aggregation method are proposed. Pooling aggregation uses global average pooling to assign the same aggregation weight to different positions, whereas weighted aggregation uses a spatial attention mechanism to focus on foreground areas in input images. A semantic-guided decoding method is further proposed that uses the output of a language model to reweight the primitive representations. Semantic-guided decoding can improve model performance on low-quality images by exploiting both visual and semantic information.

References

1. Bahdanau, D., Cho, K., Bengio, Y.: Neural machine translation by jointly learning to align and translate. In: ICLR (2015)
2. Bai, F., Cheng, Z., Niu, Y., et al.: Edit probability for scene text recognition. In: CVPR, pp. 1508–1516 (2018)
3. Cai, J., Peng, L., Tang, Y., Liu, C., Li, P.: TH-GAN: generative adversarial network based transfer learning for historical Chinese character recognition. In: ICDAR, pp. 178–183 (2019)
4. Cheng, Z., Bai, F., Xu, Y., et al.: Focusing attention: towards accurate text recognition in natural images. In: ICCV, pp. 5076–5084 (2017)
5. Choi, C., Yoon, Y., Lee, J., et al.: Simultaneous recognition of horizontal and vertical text in natural images. In: ACCV, pp. 202–212 (2018)
6. Deng, J., Dong, W., Socher, R., Li, L.J., Li, K., Li, F.F.: ImageNet: a large-scale hierarchical image database. In: CVPR, pp. 248–255 (2009)
7. Devlin, J., Chang, M.W., Lee, K., Toutanova, K.: BERT: pre-training of deep bidirectional transformers for language understanding. In: NAACL-HLT, pp. 4171–4186 (2019)
8. Fang, S., Xie, H., Wang, Y., Mao, Z., Zhang, Y.: Read like humans: autonomous, bidirectional and iterative language modeling for scene text recognition. In: CVPR, pp. 7098–7107 (2021)
9. Graves, A., Fernández, S., Gomez, F., Schmidhuber, J.: Connectionist temporal classification: labelling unsegmented sequence data with recurrent neural networks. In: ICML, pp. 369–376 (2006)
10. Gupta, A., Vedaldi, A., Zisserman, A.: Synthetic data for text localisation in natural images. In: CVPR, pp. 2315–2324 (2016)
11. He, K., Zhang, X., Ren, S., Sun, J.: Deep residual learning for image recognition. In: CVPR, pp. 770–778 (2016)
12. He, P., Huang, W., Qiao, Y., Loy, C.C., Tang, X.: Reading scene text in deep convolutional sequences. In: AAAI, pp. 3501–3508 (2016)
13. Hu, J., Shen, L., Sun, G.: Squeeze-and-excitation networks. In: CVPR, pp. 7132–7141 (2018)
14. Hu, W., Cai, X., Hou, J., et al.: GTC: guided training of CTC towards efficient and accurate scene text recognition. In: AAAI, pp. 11005–11012 (2020)

15. Huang, G., Liu, Z., Van Der Maaten, L., Weinberger, K.Q.: Densely connected convolutional networks. In: CVPR, pp. 4700–4708 (2017)
16. Jaderberg, M., Simonyan, K., Vedaldi, A., et al.: Synthetic data and artificial neural networks for natural scene text recognition. In: NIPS Workshop on Deep Learning (2014)
17. Karatzas, D., Shafait, F., Uchida, S., et al.: ICDAR 2013 robust reading competition. In: ICDAR, pp. 1484–1493 (2013)
18. Karatzas, D., Gomez-Bigorda, L., Nicolaou, A., et al.: ICDAR 2015 competition on robust reading. In: ICDAR, pp. 1156–1160 (2015)
19. Lee, C.Y., Osindero, S.: Recursive recurrent nets with attention modeling for OCR in the wild. In: CVPR, pp. 2231–2239 (2016)
20. Lin, M., Chen, Q., Yan, S.: Network in network. In: ICLR (2014)
21. Litman, R., Anschel, O., Tsiper, S., et al.: SCATTER: selective context attentional scene text recognizer. In: CVPR, pp. 11962–11972 (2020)
22. Liu, W., Chen, C., Wong, K.Y.K., Su, Z., Han, J.: STAR-Net: a spatial attention residue network for scene text recognition. In: BMVC (2016)
23. Liu, Z., Li, Y., Ren, F., et al.: Squeezedtext: a real-time scene text recognition by binary convolutional encoder-decoder network. In: AAAI, pp. 7194–7201 (2018)
24. Liu, Y., Ott, M., Goyal, N., Du, J., Joshi, M., Chen, D., Levy, O., Lewis, M., Zettlemoyer, L., Stoyanov, V.: RoBERTa: a robustly optimized BERT pretraining approach. arXiv preprint arXiv:1907.11692 (2019)
25. Long, S., He, X., Yao, C.: Scene text detection and recognition: the deep learning era. Int. J. Comput. Vis. **129**(1), 161–184 (2020)
26. Loshchilov, I., Hutter, F.: Decoupled weight decay regularization. In: ICLR (2017)
27. Lucas, S.M., Panaretos, A., Sosa, L., et al.: ICDAR 2003 robust reading competitions: entries, results, and future directions. IJDAR **7**(2), 105–122 (2005)
28. Mishra, A., Alahari, K., Jawahar, C.: Scene text recognition using higher order language priors. In: BMVC (2012)
29. Pechwitz, M., Maddouri, S.S., Märgner, V., Ellouze, N., Amiri, H., et al.: IFN/ENIT-database of handwritten Arabic words. In: CIFED, pp. 127–136 (2002)
30. Phan, T.Q., Shivakumara, P., Tian, S., et al.: Recognizing text with perspective distortion in natural scenes. In: ICCV, pp. 569–576 (2013)
31. Qiao, Z., Zhou, Y., Yang, D., et al.: SEED: semantics enhanced encoder-decoder framework for scene text recognition. In: CVPR, pp. 13528–13537 (2020)
32. Ramachandran, P., Zoph, B., Le, Q.V.: Searching for activation functions. In: NIPS, pp. 4939–4948 (2017)
33. Risnumawan, A., Shivakumara, P., Chan, C.S., et al.: A robust arbitrary text detection system for natural scene images. Expert Syst. Appl. **41**(18), 8027–8048 (2014)
34. Sheng, F., Chen, Z., Xu, B.: NRTR: a no-recurrence sequence-to-sequence model for scene text recognition. In: ICDAR, pp. 781–786 (2019)
35. Shi, B., Bai, X., Yao, C.: An end-to-end trainable neural network for image-based sequence recognition and its application to scene text recognition. IEEE Trans. Pattern Anal. Mach. Intell. **39**(11), 2298–2304 (2016)
36. Shi, B., Yao, C., Liao, M., Yang, M., Xu, P., Cui, L., Belongie, S., Lu, S., Bai, X.: ICDAR2017 competition on reading Chinese text in the wild (RCTW-17). In: ICDAR, pp. 1429–1434 (2017)
37. Shi, B., Yang, M., Wang, X., et al.: ASTER: an attentional scene text recognizer with flexible rectification. IEEE Trans. Pattern Anal. Mach. Intell. **41**(9), 2035–2048 (2019)
38. Su, B., Lu, S.: Accurate scene text recognition based on recurrent neural network. In: ACCV, pp. 35–48 (2014)
39. Tan, M., Le, Q.V.: EfficientNet: rethinking model scaling for convolutional neural networks. In: ICML, pp. 6105–6114 (2019)
40. Tan, M., Chen, B., Pang, R., et al.: MNASNET: platform-aware neural architecture search for mobile. In: CVPR, pp. 2820–2828 (2019)

41. Vaswani, A., Shazeer, N., Parmar, N., et al.: Attention is all you need. In: NIPS, pp. 5998–6008 (2017)
42. Wang, K., Babenko, B., Belongie, S.: End-to-end scene text recognition. In: ICCV, pp. 1457–1464 (2011)
43. Wang, T., Zhu, Y., Jin, L., Luo, C., Chen, X., Wu, Y., Wang, Q., Cai, M.: Decoupled attention network for text recognition. In: AAAI, pp. 12216–12224 (2020)
44. Yan, R., Peng, L., Xiao, S., Yao, G.: Primitive representation learning for scene text recognition. In: CVPR, pp. 284–293 (2021)
45. Yang, Z., Dai, Z., Yang, Y., Carbonell, J., Salakhutdinov, R.R., Le, Q.V.: XLNet: generalized autoregressive pretraining for language understanding. In: NeurIPS, pp. 5753–5763 (2019)
46. Yue, X., Kuang, Z., Lin, C., et al.: RobustScanner: dynamically enhancing positional clues for robust text recognition. In: ECCV, pp. 135–151 (2020)
47. Zeiler, M.D.: ADADELTA: an adaptive learning rate method. arXiv preprint arXiv:1212.5701 (2012)

Chapter 3
Multielement Attention Mechanism

Abstract This chapter proposes a multielement attention (MEA) mechanism, which is a generalized form of the self-attention mechanism. MEA models the spatial dependencies of input images by modeling elements in feature maps as nodes of a graph and using adjacency matrices to aggregate information from adjacent nodes when calculating attention weights. By designing different adjacency matrices, this chapter proposes three types of MEAs to learn local, neighboring, and global spatial information. An approach to integrate primitive representation learning and a multielement attention mechanism is also proposed by using visual text representations to provide global visual guidance for the decoding process. The experimental results show that MEAs that learn neighboring and global information can achieve better results than the self-attention mechanism can and that incorporating primitive representation learning can further improve model performance.

Keywords Multielement attention mechanism · Adjacency matrices · Encoder-decoder · Irregular scene text samples

3.1 Introduction

Although the performance of scene text image recognition models has significantly improved in recent years, owing to the wide variances in the contents, styles, orientations, and image quality of scene text instances, one essential research problem is how to find intrinsic representations for scene texts.

As described in Sect. 1.3.2, with the emergence of deep learning, scene text recognition is usually treated as a sequence-to-sequence modeling task. RNNs with gating mechanisms, such as LSTM [13] or GRU [7], have been widely used for temporal dependency modeling. For the decoding process, connectionist temporal classification (CTC) [9] and encoder-decoder-based frameworks with an attention mechanism [6, 16] have shown promising results. However, most existing methods transform 2D images into 1D feature sequences in the feature extraction process, which limits the model performance on irregular text images.

The Transformer model, which is based on the self-attention mechanism, has achieved superior results in various natural language processing tasks [31]. To better recognize irregular text images, some researchers have proposed the use of the powerful sequence modeling capability of the Transformer model. Sheng et al. [26] first apply the Transformer model to the scene text recognition task, but the method still follows the process of first converting the feature map into 1D feature sequences and then using the Transformer model for sequence modeling. Lyu et al. [21] proposed a 2D attention mechanism that uses a self-attention mechanism to encode features on 2D feature maps.

This chapter further explores how to learn the intrinsic representation of scene texts. To exploit the geometric structure information of scene text images, one possible way is to use graph neural networks [35]. This chapter proposes a multielement attention (MEA) mechanism, which is a generalized form of the self-attention mechanism. The elements in feature maps, which correspond to subblocks in an input image, are taken as the nodes of an undirected graph. In this way, correlations among elements in feature maps can be represented by an adjacency matrix. "Multielement" means that multiple feature elements on feature maps are utilized in different ways when computing attention weights.

Previous studies [11, 22] have shown that human visual perception includes both local and global analyses. Therefore, we propose three types of MEAs with different adjacency matrices to exploit the geometric structures of scene texts at the local, neighbor, and global levels, as shown in Fig. 3.1. If only the local adjacency matrix is used, the MEA mechanism degenerates to the self-attention form. MEA provides

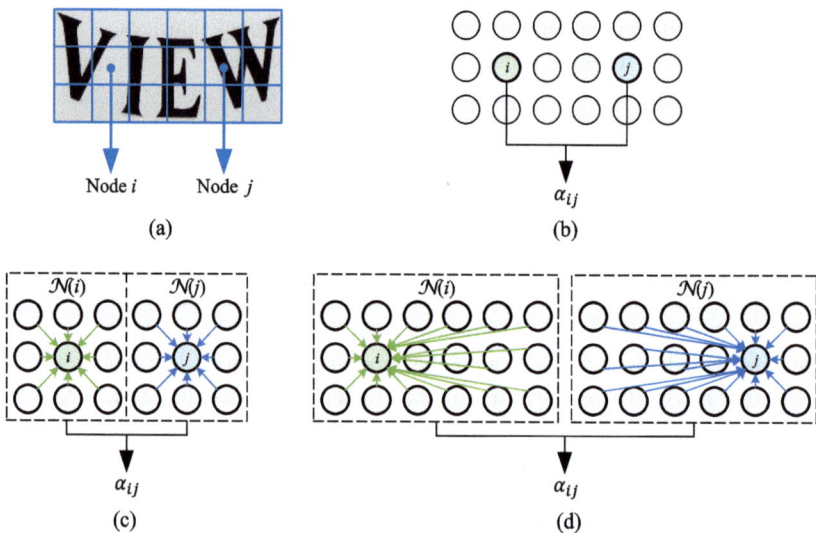

Fig. 3.1 Attention weight computations in MEA-Local, MEA-Neighbor, and MEA-Global. (**a**) Graph modeling of an input image. (**b**) MEA-Local. (**c**) MEA-Neighbor. (**d**) MEA-Global

3.2 Multielement Attention Mechanism

an alternative way to model 2D image spatial dependencies to learn the intrinsic geometric structures of scene text images with different layouts.

A multielement attention network (MEAN) is further implemented, which includes a feature extraction module, an encoder with an MEA mechanism and a decoder for predicting text codes. The feature extraction module of the MEAN is a fully convolutional network. Orientational positional encoding is added to the feature map output by the feature extraction module, which injects both orientational and positional information into the feature map. The feature maps are transformed into a feature vector sequence. The encoder-decoder module can handle feature vector sequences extracted from horizontal, vertical and irregular text images. For example, it is possible to process an upright vertical scene text sample directly without extra and improper rotation.

Finally, the integration of the MEA mechanism and the primitive representation learning proposed in the previous chapter is further explored. This chapter proposes combining visual text representations learned by the PREN with character embedding vectors as the input of the decoder. In this way, visual text representations can provide global visual guidance for the decoding process.

The experimental results of English and Chinese scene text image recognition show that the model incorporating the MEA mechanism can achieve better performance than the model based on the self-attention mechanism can, and the model combining the MEA mechanism and primitive representation learning can further achieve higher recognition accuracy.

The main contributions of this chapter are as follows:

1. A multielement attention mechanism is proposed to learn spatial dependencies by aggregating adjacent features from local to global levels before computing attention weights.
2. A multielement attention network with an orientation positional encoding strategy is implemented for the recognition of horizontal, vertical and irregular scene text samples in English and Chinese scripts.
3. This chapter proposes a method that combines the MEA mechanism and primitive representation learning.

This chapter first introduces the MEA mechanism and the implementation of three different types of MEAs and then presents the network structure of the MEAN and the method of integrating primitive representation learning. Finally, the experimental results on both English and Chinese scene text datasets are presented. The main content of this chapter comes from the authors' published paper [36].

3.2 Multielement Attention Mechanism

The MEA mechanism is a generalized form of the self-attention mechanism.

The attention mechanism is widely applied in sequence-to-sequence tasks, including text recognition. For an RNN decoder with an attention mechanism, a feedforward layer is used to learn attention weights in each decoding time step. The output of the encoder is subsequently weighted and summed via the attention weights, and the result is used as the input in the decoder RNN. Finally, a fully connected layer classifies characters from the output of the decoder. The decoding process continues until the model predicts a symbol that indicates the end of decoding or until the decoding length reaches a preset maximum value.

$$e_{t,i} = \boldsymbol{w}^T \tanh\left(W\boldsymbol{s}_{t-1} + V\boldsymbol{x}_i + \boldsymbol{b}\right) \tag{3.1}$$

$$\alpha_{t,i} = \mathrm{softmax}(e_{t,i}) \tag{3.2}$$

$$\boldsymbol{g}_t = \sum_{i=1}^{m} \alpha_{t,i} \boldsymbol{x}_i \tag{3.3}$$

$$(\boldsymbol{h}_t, \boldsymbol{s}_t) = \mathrm{rnn}\left(\boldsymbol{s}_{t-1}, [\boldsymbol{g}_t; f(y_{t-1})]\right) \tag{3.4}$$

$$p(y_t) = \mathrm{softmax}\left(W_o \boldsymbol{h}_t + \boldsymbol{b}_o\right) \tag{3.5}$$

Equations (3.1)–(3.5) describe the calculation process of an attentional RNN decoder, where $\boldsymbol{x}_i$ is the i-th element of the encoder output and where m is the length of the encoder output. $e_{t,i}$ and $\alpha_{t,i}$ are attention weights before and after Softmax normalization, respectively. $\boldsymbol{g}_t$ is the weighted summation of the encoder output and is called a "glimpse" in some studies. $\boldsymbol{h}_t$ and $\boldsymbol{s}_t$ are the hidden state and cell state of the RNN at the t-th time step, respectively. $f(y_{t-1})$ refers to the character embedding vector of the character decoded in the last time step. $[\cdot;\cdot]$ denotes concatenation. $\boldsymbol{w}, W, V, W_o, \boldsymbol{b}$, and $\boldsymbol{b}_o$ are learnable parameters. $\mathrm{rnn}(\cdot)$ refers to the mapping function of an RNN. Usually, the RNN in the decoder adopts an LSTM [13] or a gated recurrent unit (GRU) [7].

In addition to using RNNs, the Transformer model can also be applied to text recognition tasks [26]. Compared with RNNs, Transformers can better model long-term dependencies [2] through self-attention and position encoding technologies.

The self-attention mechanism can be formulated as

$$\boldsymbol{\alpha} = \mathrm{softmax}\left(\frac{1}{\sqrt{d}} X W_Q W_K^T X^T\right) \tag{3.6}$$

$$Z = \boldsymbol{\alpha} X W_V \tag{3.7}$$

where $X, Z \in \mathbb{R}^{m \times d}$ are the input and output feature vector sequences of the self-attention module. m is the number of elements in a feature vector sequence, and d is the dimension of the element in feature vector sequences. $W_Q, W_K, W_V \in \mathbb{R}^{d \times d}$ are three learnable matrices with respect to queries, keys and values. The attention weights are denoted by $\boldsymbol{\alpha} = \{\alpha_{ij}\}, i, j = 1, 2, \ldots, m$, where α_{ij} can be regarded as the correlation between elements i and j in a feature vector sequence.

The Transformer decoder predicts the next character according to both the previous output of the decoder and the encoder output. The Transformer decoder usually consists of a self-attention layer and a cross-attention layer. The self-

3.2 Multielement Attention Mechanism

attention layer aims to learn semantic dependencies from previously decoded characters, whereas the cross-attention layer learns the alignments between features and characters. The decoder runs differently during the training and inference stages. In the training stage, a technique called "teacher forcing" is utilized by feeding the corresponding character in the ground-truth text instead of the predicted output character from the previous time step as input to the current time step. This is implemented by masking out (setting to negative infinity) all future characters for the current time step from the ground-truth text embeddings in the calculation of the self-attention mechanism of the decoder [31]. Thus, the model can be trained in parallel. In the inference stage, ground-truth text is not available for the decoding process. Thus, the predicted output from the previous time step is fed as input to the current time step.

As shown in Eqs. (3.6) and (3.7), the calculation process of the self-attention mechanism can be divided into two steps: first, the attention weights α are computed by accessing each feature pair (x_i, x_j), $i, j = 1, 2, \ldots, m$; second, a linear transformation is applied to the weighted features. The first step in calculating the attention weights is the core step of the self-attention mechanism. In this process, each α_{ij} involves only x_i and x_j, whereas the correlations of other elements are neglected.

However, scene text images usually have complex 2D spatial correlations. When learning the dependency between two elements, it can be beneficial to consider their adjacent elements jointly. On the basis of the above analysis, a multielement attention (MEA) mechanism is proposed to better exploit the 2D geometric structures of scene texts. An undirected graph $\mathcal{G}$ is constructed, where the elements in feature maps extracted from an input image are taken as the nodes of $\mathcal{G}$, as shown in Fig. 3.1a. During the computation of the attention weight α_{ij}, the neighborhoods of node i and node j are aggregated according to

$$\alpha_{ij} = \text{softmax}\left(\frac{1}{\sqrt{d}} f\left(\mathcal{N}(x_i)\right) W_Q W_{rmK}^T g\left(\mathcal{N}(x_j)\right)^T\right) \quad (i, j = 1, 2, \ldots, m) \tag{3.8}$$

where $\mathcal{N}(x_i)$ and $\mathcal{N}(x_j)$ denote the neighbor nodes of node i and node j, respectively. $f(\cdot)$ and $g(\cdot)$ are two mapping functions for aggregating the features of $\mathcal{N}(x_i)$ and $\mathcal{N}(x_j)$. Before the attention weights are calculated, the MEA first fuses the information in the neighboring nodes.

In this way, the calculation of the attention weights can be regarded as the process of calculating the distance between two subgraphs. The aggregation of adjacent nodes is similar to using the mean value of each subgraph to measure the distance, which may be more stable than using only a single node.

This chapter proposes the use of the adjacency matrix of graph $\mathcal{G}$ to aggregate adjacent nodes. Let $A, B \in \mathbb{R}^{m \times m}$ be two adjacency matrices corresponding to the queries and keys; the calculation process of the MEA can be formulated as

$$\text{MEA}(X) = \text{softmax}\left(\frac{1}{\sqrt{d}} A X W_Q W_K^T X^T B^T\right) X W_V \tag{3.9}$$

By assigning different constraints to A and B, we can utilize various elements in the feature vector sequence X to compute an attention weight α_{ij}. Three kinds of adjacency matrices for the MEA are designed to aggregate information at the local, neighborhood and global levels.

1. **MEA-Local**: $A = B = I$ are identity matrices. AXW_Q and $W_K^T X^T B^T = (BXW)^T$ can be implemented as 1×1 convolution kernels. Under this constraint, MEA-Local is equivalent to the self-attention mechanism.
2. **MEA-Neighbor**: $A_{ik} = 1$ if element i and element k are spatially adjacent in the original feature maps, where $A_{ik} = 0$ and $B = A$. AXW_Q and $W_K^T X^T B^T = (BXW)^T$ can be implemented as $m \times n$ convolutions (e.g., 3×3) on the original feature maps. Under this constraint, $\mathcal{G}$ is a connected graph, and elements centered on positions i and j are used in computing α_{ij}.
3. **MEA-Global**: A and B are randomly initialized and learned in the training stage. AXW_Q and $W_K^T X^T B^T = (BXW)^T$ can be implemented as graph convolutional layers [5, 15]. Under this constraint, $\mathcal{G}$ is a complete graph, and all the elements are used to compute α_{ij}.

3.3 Multielement Attention Network

Figure 3.2 illustrates the framework of the MEAN. A U-shaped CNN first extracts feature maps from the input image. The orientational positional encoding is subsequently added to the feature maps. An encoder with the MEA mechanism encodes the features into hidden representations. Finally, a decoder progressively generates predictions from the hidden representations. The detailed structure of the MEAN is described below.

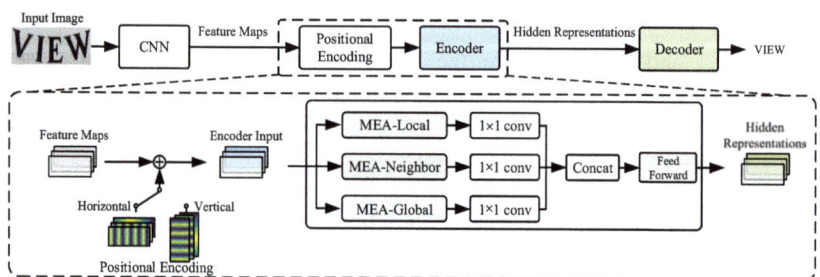

Fig. 3.2 The system framework of the multielement attention network

3.3 Multielement Attention Network

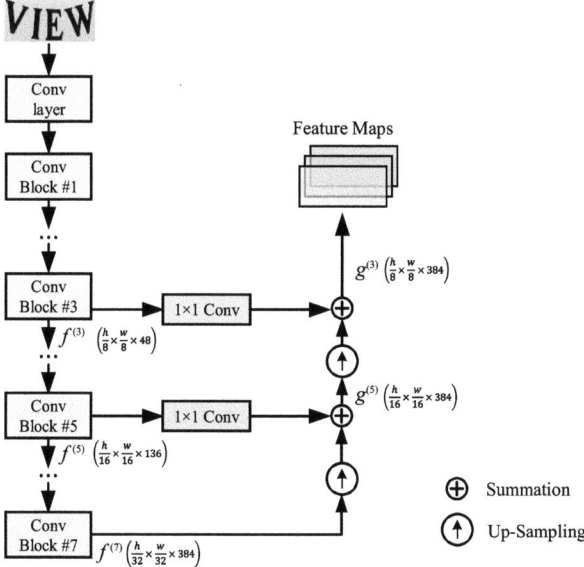

Fig. 3.3 The structure of the CNN used in the multielement attention network

3.3.1 Feature Extraction

The feature extraction module of the MEAN is a CNN based on EfficientNet-B3 [30]. To exploit multiscale feature maps, MEAN adopts a U-shaped CNN structure [25]. As shown in Fig. 3.3, let the feature map output by the l-th convolutional block in EfficientNet-B3 be $f^{(l)}$. The output of the final block $f^{(7)}$ is upsampled and summed with $f^{(5)}$ to form new feature maps $g^{(5)}$; then, $g^{(5)}$ is upsampled again and summed with $f^{(3)}$ to obtain $g^{(3)}$, which is used as the output of the feature extraction module. The dimensions of the output feature maps are $\frac{h}{8} \times \frac{w}{8} \times d$, where h and w are the height and width of the input image, respectively, and where d is the number of channels. In this way, the final features contain both local information from the bottom layers and semantic information from the top layers, and can produce better representations for characters of varying sizes.

3.3.2 Orientational Positional Encoding

The original positional encoding (PE) [31] is used for neural machine translation where the inputs are 1D sequences. To better handle multioriented texts, we propose adding orientational positional information into feature maps according to the horizontal and vertical reading directions. The proposed orientational positional encoding is calculated as

$$PE^{(H)}_{(i,j,2k)} = \sin\left(j/L^{2k/d}\right) \qquad (3.10)$$

$$PE^{(H)}_{(i,j,2k+1)} = \cos\left(j/L^{2k/d}\right) \qquad (3.11)$$

$$PE^{(V)}_{(i,j,2k)} = \sin\left(i/L^{2k/d}\right) \qquad (3.12)$$

$$PE^{(V)}_{(i,j,2k+1)} = \cos\left(i/L^{2k/d}\right) \qquad (3.13)$$

where (i, j) denotes a position on the 2D feature maps. $2k$ and $2k + 1$ are indices of the channel dimension, and d is the number of channels. H and V refer to the horizontal and vertical directions, respectively. L is a preset constant.

For horizontal input images, position encodings are generated according to the image width. For vertical input images, position encodings are generated according to the image height. The orientational positional encoding is added to the feature maps. Then, the feature maps are transformed into a feature vector sequence through flattening, which is used as the input of the subsequent encoder-decoder module.

3.3.3 Encoder and Decoder

As shown in Fig. 3.2, three levels of the MEA mechanism are computed in parallel in the encoder, and their outputs are concatenated. To facilitate the stacking of multiple encoder layers, we use 1×1 convolutions to reduce the output dimension of each MEA layer to 1/3 of the original data dimension. Therefore, the concatenated results have the same data dimension as that of the encoder's input.

The concatenated output of the MEA layers is then transformed by a feedforward network, which is calculated as

$$\text{FFD}(X) = \phi\left(XW_1 + \boldsymbol{b}_1\right) W_2 + \boldsymbol{b}_2 \qquad (3.14)$$

where $\text{FFD}(\cdot)$ is the mapping function of the feedforward network. X is the concatenated output of three MEA layers. W_1, W_2, $\boldsymbol{b}_1$, and $\boldsymbol{b}_2$ are learnable network parameters. $\phi(\cdot)$ is the GELU [12] activation function. The encoder converts feature maps with orientational positional encodings into hidden representations as the input to the decoder.

For the decoding process, MEAN adopts the Transformer decoder [31] for text transcription. To better exploit the temporal dependencies of output texts from both the forward and backward directions, two decoders with opposite decoding directions are trained simultaneously [29].

3.4 Primitive Representation Learning with Multielement Attention

In the training stage, the cross entropy between the predicted texts and the ground truth is used as the loss function, which is shown in Eq. (3.15).

$$\mathcal{L} = -\frac{1}{2(l+1)} \sum_{r=1}^{2} \sum_{t=1}^{l+1} \log p\left(y_t^{(r)} | I\right) \tag{3.15}$$

where I is the input image, y_t ($t = 1, 2, \ldots, l$) is the t-th character in the ground-truth, and l is the length of the label text. y_{l+1} is the ⟨eos⟩ symbol. r indicates the forward or backward direction of the decoder.

In the test stage, the prediction result is assigned a higher recognition score, which is defined as the average of the log-Softmax values of all the predicted characters, i.e.,

$$s^{(r)} = \frac{1}{\tilde{l}^{(r)}} \sum_{t=1}^{\tilde{l}^{(r)}} \log p\left(\tilde{y}_t^{(r)}\right) \tag{3.16}$$

where $s^{(r)}$ is the recognition score of the r direction. $\tilde{y}_t^{(r)}$ is the t-th character predicted by the decoder in the r direction. $\tilde{l}^{(r)}$ is the length of the text predicted by the decoder in the r direction. The decoding process continues until the ⟨eos⟩ symbol is predicted and when $\tilde{l}^{(r)}$ reaches the preset maximum decoding length.

3.4 Primitive Representation Learning with Multielement Attention

The multielement attention network is an MEA-based encoder-decoder model. Although the MEA mechanism is introduced in the encoding process to learn the spatial structure information of input images, the decoding process is worthy of further analysis. For example, the attention drift problem may occur when the alignment between feature maps and text is calculated because the decoding process heavily relies on previously decoded characters. Since the PREN proposed in Chap. 2 is completely based on visual information for text recognition, it can provide global visual guidance for the decoding process to learn more stable and precise alignments.

This chapter proposes a combination of multielement attention mechanisms with primitive representation learning. Since the MEA is a type of 2D attention mechanism, the model with both the MEA and primitive representation learning is named PREN2D. As shown in Fig. 3.4, both the PREN and MEAN share the same feature extraction module. Visual text representations learned by the PREN are combined with character embedding vectors in the decoder through a gated

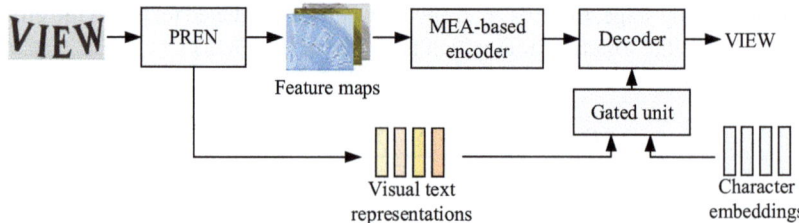

Fig. 3.4 Network structure of PREN2D, which combines primitive representation learning and a multielement attention mechanism

unit. Let V be the visual text representation; then, the updated decoder input can be calculated as

$$G = \sigma\left([V; E] \cdot W_g\right) \tag{3.17}$$

$$E' = G \odot V + (1 - G) \odot E \tag{3.18}$$

where the character embedding E is computed in two steps. First, the one-hot encodings of the text are transformed into embedding vectors by a linear layer. Here the input text is the ground truth in the training stage or the previously decoded results in the test stage. Let the dimension of the weight matrix in the linear layer be $C \times d$, where C is the number of character classes and where d is the dimension of the embedding vectors. The sinusoidal positional encodings proposed by Vaswani et al.[31] are added to the embedding vectors to obtain E. $[\cdot; \cdot]$ denotes concatenation. W_g is a learnable parameter matrix, and $\odot$ denotes the elementwise product. G is used to control the gated unit. To reduce computational complexity, PREN2D uses only a unidirectional decoder, which decodes text in reading order.

In the training stage, PREN2D uses both visual text representations and outputs of the encoder-decoder module for decoding. The objective function of the network is

$$\mathcal{L} = \lambda \mathcal{L}_{\text{pren}} + \mathcal{L}_{\text{enc-dec}} \tag{3.19}$$

where $\mathcal{L}_{\text{pren}}$ is the cross-entropy loss between the ground-truth text and the character probabilities calculated from visual text representations and where $\mathcal{L}_{\text{enc-dec}}$ is the cross-entropy loss between the ground-truth text and the character probabilities calculated from outputs of the encoder-decoder module. $\lambda = 0.5$ is a preset constant.

In the test stage, PREN2D uses the output of the encoder-decoder module as the final prediction.

3.5 Discussion

This section discusses the performance of models with different MEA mechanisms. Different position encoding strategies are also compared for the multioriented Chinese scene text recognition task. Finally, MEAN and PREN2D are compared with a series of recent scene text image recognition methods.

3.5.1 Experimental Settings

For English scene text image recognition, all the models are trained on MJSynth [14] and SynthText [10] synthetic datasets for 8 epochs with a batch size of 128. The optimizer is ADADELTA [39]. The learning rate is initialized to 0.5 and reduced by 0.2, 0.1 and 0.1 times, at the third, sixth, and eighth epochs, respectively. For MEA-neighbor, the convolutional kernel size is set to 3×3, i.e., nine centered elements are used when calculating each attention weight. The numbers of layers in both the encoder and decoder are set to two. For orientational positional encoding, the constant L in Eqs. (3.10)–(3.13) is set to 64. All the input images are normalized to 64×256 pixels.

For multioriented Chinese scene text image recognition, all the models are trained on the SynthChinese synthetic dataset for five epochs and then fine-tuned on a subset of the RCTW dataset [28] for 20 epochs. The learning rate is initialized to 1.0 and decreases to 0.1 in the sixth epoch. The horizontal text images are normalized to 64×256 pixels, and the vertical text images are normalized to 256×64 pixels.

3.5.2 Comparison of Different Types of Multielement Attention Mechanisms

Table 3.1 lists the word accuracy and numbers of parameters of the models using different MEA mechanisms in the English scene text recognition task, where "Joint Scheme" means using all three types of MEAs in the encoder. When only one

Table 3.1 Word accuracy (%) of models with different multielement attention mechanisms in the English scene text recognition task

MEA	IIIT5K	SVT	IC03	IC13	IC15	SVTP	CUTE	Average	#Params
MEA-Local	95.1	93.5	95.2	95.6	85.0	85.5	86.1	91.7	23.4 M
MEA-Neighbor	95.4	93.8	95.3	96.1	85.5	86.8	86.5	92.1	28.1 M
MEA-Global	95.3	93.5	95.2	95.1	85.2	85.9	86.8	91.8	23.7 M
Joint scheme	**95.8**	**94.3**	**95.7**	**96.6**	**85.9**	**87.8**	**87.1**	**92.6**	31.0 M

Bold values highlight the optimal performance (e.g., highest accuracy or lowest error rate) for the respective metric across different methods, models, or systems

Table 3.2 Word accuracy (%) of models with different multielement attention mechanisms in the multioriented Chinese scene text recognition task

MEA	Horizontal	Vertical	Average	#Params
MEA-Local	80.2	83.2	81.7	29.9 M
MEA-Neighbor	81.2	83.8	82.5	34.6 M
MEA-Global	80.8	84.8	82.8	30.1 M
Joint scheme	**81.6**	**86.0**	**83.8**	37.5 M

Bold values highlight the optimal performance (e.g., highest accuracy or lowest error rate) for the respective metric across different methods, models, or systems

MEA mechanism is used, the model with MEA-Neighbor achieves the highest average word accuracy. Models with MEA-Neighbor and MEA-Global achieve better performance than the model with MEA-Local does, indicating that it is beneficial to consider more feature elements in the calculation of attention weights. Moreover, the model with all three types of MEAs achieves the best performance on all test sets. Compared with the model that uses only MEA-Neighbor, the model that integrates three types of MEAs has an average word accuracy improvement of 0.5%, indicating that different MEAs are complementary.

Table 3.2 lists the performances of the models that use different MEA mechanisms in the multioriented Chinese scene text recognition task. The word accuracy of the models with MEA-Neighbor and MEA-Global can still exceed that of the model with MEA-Local, and the model with three types of MEAs can still achieve the best performance.

3.5.3 Effect of Orientational Positional Encoding

To study the effect of orientational positional encoding, models with three different positional encoding strategies are compared on the multioriented Chinese scene text recognition task. In Table 3.3, "None" means that no positional encoding is used. "Learnable" refers to positional encodings that are randomly initialized and trained with the model. Learnable positional encoding is also used by the BERT language model[8]. "Orientational" is the proposed orientational positional encoding.

Table 3.3 lists the word accuracies of the models when different positional encoding strategies are used. Compared with the model without positional encoding, the model with learnable positional encoding has slightly higher accuracy. The model with orientational positional encoding proposed in this chapter achieves the best performance. Compared with MEAN, PREN2D achieves significantly higher word accuracy on both the horizontal and vertical test sets, indicating that incorporating primitive representation learning can effectively improve model performance.

3.5 Discussion

Table 3.3 Word accuracy (%) of models with different positional encoding strategies

Model	Positional encoding	Horizontal	Vertical	Average
MEAN	None	80.2	85.8	83.0
	Learnable	80.8	85.6	83.2
	Orientational	81.6	86.0	83.8
PREN2D	Orientational	**83.0**	**88.8**	**85.9**

Bold values highlight the optimal performance (e.g., highest accuracy or lowest error rate) for the respective metric across different methods, models, or systems

3.5.4 Comparison with Existing Methods

Table 3.4 lists the performance comparisons of the PREN, MEAN, and PREN2D methods and a series of recent scene text recognition methods. In Table 3.4, "MJ" and "ST" refer to the MJSynth [14] and SynthText [10] datasets, respectively, and "Char" indicates that the character-level bounding box annotations are used in the training stage. "Add" means that additional training data are used for training. The model marked with the "*" symbol indicates that the original model is not trained on the MJSynth and SynthText datasets, whereas Baek et al. [1] reimplemented the model and trained it on the MJSynth and SynthText datasets for fair comparison.

With the integration of primitive representation learning and a multielement attention mechanism, PREN2D achieves the best performance on four regular text test sets (IIIT5K, SVT, IC03, and IC13) and two irregular text test sets (IC15 and SVTP).

3.5.5 Visualization

3.5.5.1 Visualization of Different Types of MEAs

To better understand the MEA mechanism, an additional Chinese character recognition task [4] is conducted by using a CNN-based classifier with only one MEA layer after the second convolutional block in EfficientNet-B3 [30]. Figure 3.5 visualizes the attention weights α_{ij} ($j = 1, 2, \ldots, m$), where i and j are the indices in the feature vector sequence. The corresponding position of i in the feature maps is marked by the red point (denoted as P_i) on the input image. On the one hand, the attention weights of MEA-Neighbor and MEA-Global in the background are generally less than those of MEA-Local, indicating that considering adjacent features can help the model focus on areas with more textual information. On the other hand, the attention of MEA-Neighbor is concentrated in a few foreground areas, whereas the attention of MEA-Global covers most foregrounds, indicating that different types of MEA mechanisms are complementary.

Table 3.4 Word accuracy (%) across different models and datasets

Model	Training data	IIIT5K	SVT	IC03	IC13	IC15	SVTP	CUTE
Mask TextSpotter (Liao et al. [18])	MJ + ST + Char	95.3	91.8	95.0	95.3	78.2	83.6	88.5
SAR (Li et al. [17])	MJ + ST + Add	95.0	91.2	–	94.0	78.8	86.4	89.6
SCATTER (Litman et al. [19])	MJ + ST + Add	93.7	92.7	**96.3**	93.9	82.2	86.9	87.5
CRNN (Shi et al. [1, 27])*	MJ + ST	82.9	81.6	92.6	91.1	69.4	70.0	65.5
ASTER (Shi et al. [29])	MJ + ST	93.4	89.5	94.5	91.8	76.1	78.5	79.5
MORAN (Luo et al. [20])	MJ + ST	91.2	88.3	95.0	92.4	68.8	76.1	77.4
ESIR (Zhan et al. [40])	MJ + ST	93.3	90.2	–	91.3	76.9	79.6	83.3
TextScanner (Wan et al. [32])	MJ + ST	93.9	90.1	–	92.9	79.4	84.3	83.3
DAN (Wang et al. [33])	MJ + ST	94.3	89.2	95.0	93.9	74.5	80.0	84.4
SE-ASTER (Qiao et al. [23])	MJ + ST	93.8	89.6	–	92.8	80.0	81.4	83.6
AutoSTR (Zhang et al. [41])	MJ + ST	94.7	90.9	93.3	94.2	81.8	81.7	–
SRN (Yu et al. [37])	MJ + ST	94.8	91.5	–	95.5	82.7	85.1	87.8
RobustScanner (Yue et al. [38])	MJ + ST	95.3	88.1	–	94.8	77.1	79.5	**90.3**
GA-SPIN (Zhang et al. [42])	MJ + ST	95.2	90.9	94.9	94.8	82.8	83.2	87.5
Bhunia et al. [3]	MJ + ST	95.2	92.2	–	95.5	84.0	85.7	89.7
PIMNet (Qiao et al. [24])	MJ + ST	95.2	91.2	–	95.2	83.5	84.3	84.4
VisionLAN (Wang et al. [34])	MJ + ST	95.8	91.7	–	95.7	83.7	86.0	88.5
PREN	MJ + ST	91.7	88.9	93.3	94.2	79.7	82.5	84.7
PREN (semantic)	MJ + ST	93.4	92.9	94.9	96.4	82.9	86.5	88.5
MEAN	MJ + ST	95.8	94.3	95.7	96.6	85.9	87.8	87.1
PREN2D	MJ + ST	**96.5**	**95.1**	96.3	**96.8**	**86.0**	**88.1**	88.5

Bold values highlight the optimal performance (e.g., highest accuracy or lowest error rate) for the respective metric across different methods, models, or systems

3.5 Discussion

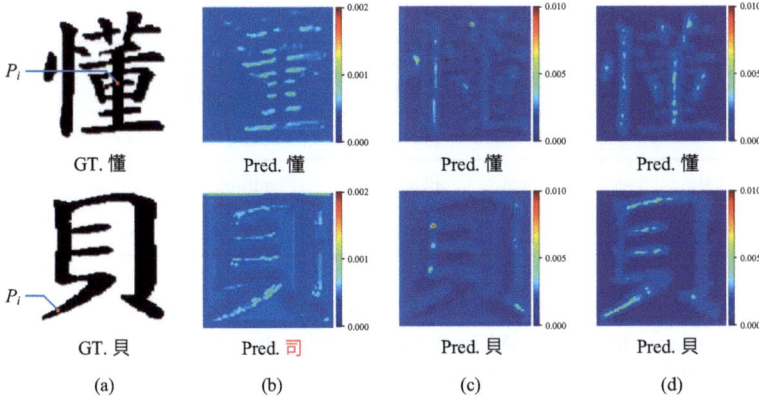

Fig. 3.5 Visualization of attention weights α_{ij} for different components of the MEA mechanism. (**a**) Input image, where the red point denotes position i. (**b**)–(**d**) The attention weights of MEA-Local, MEA-Neighbor, and MEA-Global, respectively

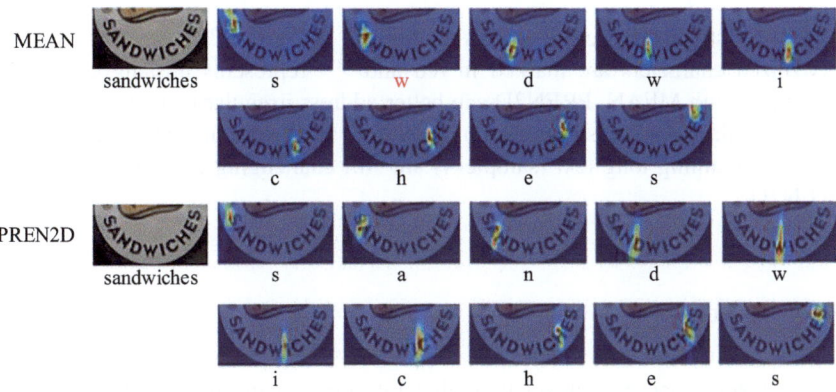

Fig. 3.6 Visualization of attention weights generated in the decoding process before and after incorporating primitive representation learning

3.5.5.2 Effect of Incorporating Primitive Representation Learning

The results in Tables 3.3 and 3.4 demonstrate that after incorporating primitive representation learning, PREN2D can achieve better performance than MEAN. Figure 3.6 shows the attention weights calculated by the decoders of MEAN and PREN2D with respect to the same input image. The input image is listed in the first column, followed by the visualization results of the attention weights for each decoding time step. The predicted characters are listed below the attention weight heatmaps, and the incorrectly recognized characters are marked in red. By combining visual text representations, PREN2D can generate more stable and accurate attention weights, thus avoiding some incorrect alignments. For example,

Table 3.5 Recognition examples of MEAN and PREN2D

Input image				
Ground-truth text	unicorn	pharmacy	londitioned	yankeecandli
MEAN	canicoes	pharmagy	conditioned	yankeecandu_
PREN2D	unicorn	pharmacy	conditioned	yankeecandle

MEAN confuses the letter "A" in the image with the letter "W", and misses the recognition of the letter "N" in the image. In contrast, PREN2D can avoid such recognition errors.

3.5.5.3 Error Analysis

Table 3.5 lists recognition examples of MEAN and PREN2D, where incorrectly recognized characters are marked in red and "_" represents missing characters. Compared with MEAN, PREN2D can better address irregular text layouts (sample 1) and similar characters (sample 2). However, some special fonts (sample 3) or images containing long text (sample 4) are still challenging for both MEAN and PREN2D.

3.6 Summary

This chapter proposes a novel multielement attention mechanism for scene text recognition. MEA models the spatial dependencies of input images by modeling elements in feature maps as nodes of a graph and using adjacency matrices to aggregate information from adjacent nodes when calculating attention weights. By designing different adjacency matrices, this chapter proposes three types of MEAs to learn local, neighboring, and global spatial information. An approach to integrate primitive representation learning and a multielement attention mechanism is also proposed by using visual text representations to provide global visual guidance for the decoding process. The experimental results show that MEAs that learn neighboring and global information can achieve better results than the self-attention mechanism can and that incorporating primitive representation learning can further improve model performance.

References

1. Baek, J., Kim, G., Lee, J., et al.: What is wrong with scene text recognition model comparisons? Dataset and model analysis. In: ICCV, pp. 4715–4723 (2019)
2. Bengio, Y., Simard, P., Frasconi, P.: Learning long-term dependencies with gradient descent is difficult. IEEE Trans. Neural Netw. **5**(2), 157–166 (1994)
3. Bhunia, A.K., Sain, A., Kumar, A., Ghose, S., Chowdhury, P.N., Song, Y.Z.: Joint visual semantic reasoning: multi-stage decoder for text recognition. In: ICCV, pp. 14940–14949 (2021)
4. Cai, J., Peng, L., Tang, Y., Liu, C., Li, P.: TH-GAN: generative adversarial network based transfer learning for historical Chinese character recognition. In: ICDAR, pp. 178–183 (2019)
5. Chen, Y., Rohrbach, M., Yan, Z., et al.: Graph-based global reasoning networks. In: CVPR, pp. 433–442 (2019)
6. Cheng, Z., Bai, F., Xu, Y., et al.: Focusing attention: towards accurate text recognition in natural images. In: ICCV, pp. 5076–5084 (2017)
7. Cho, K., van Merriënboer, B., Bahdanau, D., Bengio, Y.: On the properties of neural machine translation: encoder–decoder approaches. In: Eighth Workshop on Syntax, Semantics and Structure in Statistical Translation, pp. 103–111 (2014)
8. Devlin, J., Chang, M.W., Lee, K., Toutanova, K.: BERT: pre-training of deep bidirectional transformers for language understanding. In: NAACL-HLT, pp. 4171–4186 (2019)
9. Graves, A., Fernández, S., Gomez, F., Schmidhuber, J.: Connectionist temporal classification: labelling unsegmented sequence data with recurrent neural networks. In: ICML, pp. 369–376 (2006)
10. Gupta, A., Vedaldi, A., Zisserman, A.: Synthetic data for text localisation in natural images. In: CVPR, pp. 2315–2324 (2016)
11. Hegdé, J.: Time course of visual perception: coarse-to-fine processing and beyond. Prog. Neurobiol. **84**(4), 405–439 (2008)
12. Hendrycks, D., Gimpel, K.: Gaussian error linear units (GELUs). arXiv preprint arXiv:1606.08415 (2016)
13. Hochreiter, S., Schmidhuber, J.: Long short-term memory. Neural Comput. **9**(8), 1735–1780 (1997)
14. Jaderberg, M., Simonyan, K., Vedaldi, A., et al.: Synthetic data and artificial neural networks for natural scene text recognition. In: NIPS Workshop on Deep Learning (2014)
15. Kipf, T.N., Welling, M.: Semi-supervised classification with graph convolutional networks. In: ICLR (2017)
16. Lee, C.Y., Osindero, S.: Recursive recurrent nets with attention modeling for OCR in the wild. In: CVPR, pp. 2231–2239 (2016)
17. Li, H., Wang, P., Shen, C., Zhang, G.: Show, attend and read: a simple and strong baseline for irregular text recognition. In: AAAI, pp. 8610–8617 (2019)
18. Liao, M., Lyu, P., He, M., et al.: Mask TextSpotter: an end-to-end trainable neural network for spotting text with arbitrary shapes. IEEE Trans. Pattern Anal. Mach. Intell. **43**(2), 532–548 (2021)
19. Litman, R., Anschel, O., Tsiper, S., et al.: SCATTER: selective context attentional scene text recognizer. In: CVPR, pp. 11962–11972 (2020)
20. Luo, C., Jin, L., Sun, Z.: MORAN: a multi-object rectified attention network for scene text recognition. Pattern Recognit. **90**(1), 109–118 (2019)
21. Lyu, P., Yang, Z., Leng, X., et al.: 2D attentional irregular scene text recognizer. arXiv preprint arXiv:1906.05708 (2019)
22. Navon, D.: Forest before trees: The precedence of global features in visual perception. Cogn. Psychol. **9**(3), 353–383 (1977)
23. Qiao, Z., Zhou, Y., Yang, D., et al.: SEED: semantics enhanced encoder-decoder framework for scene text recognition. In: CVPR, pp. 13528–13537 (2020)

24. Qiao, Z., Zhou, Y., Wei, J., Wang, W., Zhang, Y., Jiang, N., Wang, H., Wang, W.: PIMNet: a parallel, iterative and mimicking network for scene text recognition. In: ACM MM, pp. 2046–2055 (2021)
25. Ronneberger, O., Fischer, P., Brox, T.: U-net: convolutional networks for biomedical image segmentation. In: MICCAI, pp. 234–241 (2015)
26. Sheng, F., Chen, Z., Xu, B.: NRTR: a no-recurrence sequence-to-sequence model for scene text recognition. In: ICDAR, pp. 781–786 (2019)
27. Shi, B., Bai, X., Yao, C.: An end-to-end trainable neural network for image-based sequence recognition and its application to scene text recognition. IEEE Trans. Pattern Anal. Mach. Intell. **39**(11), 2298–2304 (2016)
28. Shi, B., Yao, C., Liao, M., Yang, M., Xu, P., Cui, L., Belongie, S., Lu, S., Bai, X.: ICDAR2017 competition on reading Chinese text in the wild (RCTW-17). In: ICDAR, pp. 1429–1434 (2017)
29. Shi, B., Yang, M., Wang, X., et al.: ASTER: an attentional scene text recognizer with flexible rectification. IEEE Trans. Pattern Anal. Mach. Intell. **41**(9), 2035–2048 (2019)
30. Tan, M., Le, Q.V.: EfficientNet: rethinking model scaling for convolutional neural networks. In: ICML, pp. 6105–6114 (2019)
31. Vaswani, A., Shazeer, N., Parmar, N., et al.: Attention is all you need. In: NIPS, pp. 5998–6008 (2017)
32. Wan, Z., He, M., Chen, H., Bai, X., Yao, C.: TextScanner: reading characters in order for robust scene text recognition. In: AAAI, pp. 12120–12127 (2020)
33. Wang, T., Zhu, Y., Jin, L., Luo, C., Chen, X., Wu, Y., Wang, Q., Cai, M.: Decoupled attention network for text recognition. In: AAAI, pp. 12216–12224 (2020)
34. Wang, Y., Xie, H., Fang, S., Wang, J., Zhu, S., Zhang, Y.: From two to one: a new scene text recognizer with visual language modeling network. In: ICCV, pp. 14194–14203 (2021)
35. Wu, Z., Pan, S., Chen, F., Long, G., Zhang, C., Philip, S.Y.: A comprehensive survey on graph neural networks. IEEE Trans. Neural Netw. Learn. Syst. **32**(1), 4–24 (2020)
36. Yan, R., Peng, L., Xiao, S., Yao, G., Min, J.: MEAN: multi-element attention network for scene text recognition. In: ICPR, pp. 6850–6857 (2020)
37. Yu, D., Li, X., Zhang, C., et al.: Towards accurate scene text recognition with semantic reasoning networks. In: CVPR, pp. 12113–12122 (2020)
38. Yue, X., Kuang, Z., Lin, C., et al.: RobustScanner: dynamically enhancing positional clues for robust text recognition. In: ECCV, pp. 135–151 (2020)
39. Zeiler, M.D.: ADADELTA: an adaptive learning rate method. arXiv preprint arXiv:1212.5701 (2012)
40. Zhan, F., Lu, S.: ESIR: end-to-end scene text recognition via iterative image rectification. In: CVPR, pp. 2059–2068 (2019)
41. Zhang, H., Yao, Q., Yang, M., et al.: AutoSTR: efficient backbone search for scene text recognition. In: ECCV, pp. 751–767 (2020)
42. Zhang, C., Xu, Y., Cheng, Z., Pu, S., Niu, Y., Wu, F., Zou, F.: SPIN: structure-preserving inner offset network for scene text recognition. In: AAAI, pp. 3305–3314 (2021)

Chapter 4
Dynamic Temporal Residual Learning and Attention Rectification

Abstract This chapter further explores how to model long-term contextual dependencies in text images. A dynamic temporal residual learning mechanism is proposed to model the contextual information in feature sequences by introducing the residual learning method into the temporal dimension of an RNN encoder. An attention rectification method is also proposed to mitigate the attention drift problem in the decoder.

Keywords Long short-term memory · Residual learning · Handwriting recognition · Attention rectification

4.1 Introduction

In previous chapters, this book introduces primitive representation learning and multielement attention mechanisms and constructs the PREN, MEAN, and PREN2D models. These models achieve promising performance in scene text recognition tasks. However, recognizing images containing long texts is still a challenging problem. This chapter further explores how to model long-term contextual dependencies in text images.

A text image can be viewed as a temporal sequence using a sliding window. Because temporal sequences are neither always uniformly structured nor uniformly predictable [4], capturing dynamic dependencies in sequence modeling tasks such as handwriting and speech recognition is challenging. For a network with an encoder-decoder structure, the encoder models contextual dependencies in extracted features, and the decoder learns the semantic information in the text.

This chapter first proposes a dynamic temporal residual learning mechanism to model the contextual information in feature sequences by introducing the residual learning method into the temporal dimension of an RNN encoder. For text images, the temporal dimension can be regarded as the reading order of the text. Then, for the attentional decoder, this chapter proposes an attention rectification method. By introducing a parallel attention mechanism combined with character position information, the model learns the offset of attention weights for rectification, and, mitigates the attention drift problem.

The encoder of a text recognition model is usually an RNN or self-attention network. Since the computational complexity of the self-attention network is proportional to the square of the length of the input feature sequence, images with long texts rapidly increase the computational complexity of the network. Moreover, recent research [8] showed that using an RNN as the encoder achieves better performance than a self-attention network in handwriting recognition tasks. Therefore, this chapter adopts the LSTM network [13] to encode feature sequences extracted by a CNN. The LSTM network is widely used in sequence modeling tasks. The LSTM network has made considerable progress in sequence modeling by introducing a gating mechanism for the vanilla RNN, which alleviates the gradient vanishing or exploding problem [12] in training through the back-propagation through time (BPTT) method. In real applications, an LSTM network not only has multiple stacked layers in a spatial structure but also has the same depth as the length of the feature sequence when unfolded in time [19], which indicates that there is a possibility for further improvements in the temporal structure to better capture dynamic temporal dependencies in sequential data.

Residual learning has been shown to be effective for deep feed-forward neural networks via models such as ResNet [11] and Highway network [27], where ResNet uses identity connections, while the Highway network further applies the gating mechanism in residual connections.

Efforts have been made to simplify RNN optimization by constructing shortcut connections between adjacent RNN layers [31, 38], which are similar to ResNet. These works add residual connections into stacked RNN layers. However, because the temporal depth of an unfolded LSTM network is usually deeper than the spatial depth of its stacked layers, it is feasible to explore the temporal residual learning mechanism for LSTM. Gui et al. [10] proposed introducing skip connections into the LSTM network to reuse previous outputs, and the skip size was determined via reinforcement learning. However, the reinforcement learning module results in considerable additional computational complexity, and skip connections use fixed weights, which are not enough to fully model the temporal dependencies of feature sequences. Yue et al. [35] proposed adding shortcuts with identity connections or gating mechanisms into the temporal dimension of vanilla RNNs. The modified RNN is similar to a simplified LSTM or GRU. While the model has a faster recognition speed, it has not achieved better accuracy than the original GRU network.

This chapter proposes a dynamic temporal residual learning architecture for LSTM that is equivalent to adding shortcut connections between the adjacent temporal outputs of LSTM units. An independent secondary network is used to generate dynamic weights of residual connections. The secondary network aims to learn contextual information in input sequences and capture the temporal dependencies. The proposed model is called a dynamic temporal residual network (DTRN).

4.1 Introduction

Table 4.1 Character error rate (%) and word error rate (%) of models with different decoding methods on the IAM English handwriting dataset and Rimes French handwriting dataset

Decoding method	IAM				Rimes			
	Val		Test		Val		Test	
	CER	WER	CER	WER	CER	WER	CER	WER
CTC	**4.21**	15.37	**5.95**	20.00	**2.26**	10.71	**2.11**	**10.22**
Attention	4.78	**15.01**	6.50	**18.67**	2.57	**10.66**	2.33	10.27

Bold values highlight the optimal performance (e.g., highest accuracy or lowest error rate) for the respective metric across different methods, models, or systems

However, for decoding methods, attention-based encoder-decoder models have achieved high performance in scene text recognition tasks. Instead of containing the limited number of words in most scene text images, handwritten text images often contain long text lines, which have richer semantic information. Therefore, it is intuitive that the implicit language model learned by an attention decoder is more suitable for handwriting recognition tasks.

However, preliminary experiments have shown the opposite results. Table 4.1 lists the performance of models with the same encoder and different decoders on the IAM English handwriting dataset [18] and the Rimes French handwriting dataset [9]. Although the attention-based encoder-decoder model achieves a lower WER on some datasets, its CER is significantly higher than that of the CTC-based model.

The reason for the results in Table 4.1 is that the attention-based encoder-decoder model easily suffers from the attention drift problem. The decoder learns the alignment relationships between feature sequences and characters, which is sensitive to previously decoded results. For images with long texts, the attention drift problem may cause serious recognition errors. For example, Fig. 1.5a shows an input text image, and Fig. 1.5b visualizes the attention weights calculated by an attentional decoder. The alignment between the feature sequence and the characters is unstable during the decoding process, resulting in some missing characters or repeatedly recognized characters. As shown in Fig. 1.5e, the attention drift problem results in a large CER.

For the CTC-based model, since each position in the feature sequence is mapped into a character or a "blank" symbol, there is a natural alignment between the feature sequence and the text, making the CTC-based model have a lower CER than the attention-based model does. However, the CTC-based model cannot exploit the semantic information contained in the text. It is challenging for CTC-based models to differentiate similar characters, so the model has a higher WER.

According to the above analysis, effectively applying an attention mechanism to the recognition of images with long text lines is an important research problem. In recent years, efforts have been made to alleviate the attention drift problem. The FAN model [1] uses a focusing network to learn the attention center of each character on feature maps. An additional objective function is introduced in the training process to minimize the difference between the predicted attention centers and the ground-truth character centers. However, training the FAN model requires character position annotations. For handwritten text images with many cursive scripts or ligatures, annotating all the character bounding boxes is difficult. The DAN model [30] adopted a fully convolutional network to predict the alignments between features and characters by using visual information. However, for low-quality images, the visual features extracted by a CNN may contain noise that makes the predicted alignments inaccurate. The RobustScanner model [36] uses a position enhancement branch, which uses a parallel attention module to learn features that contain rich position information. The output features of the parallel attention module are fused with the output features of the original decoder. The method is validated in the scene text recognition task. However, for images with long text lines, it is difficult for the parallel attention module to accurately predict the text in the input image, and the fusion of the output features of the two decoders may degrade the model performance.

To solve the above problems, this chapter proposes an attention rectification network (ARN). Unlike the FAN model [1], training the ARN does not require character-level annotations, and the proposed method can be easily applied to existing attention-based encoder-decoder models. Unlike the DAN model [30], the ARN aims to perform a simpler and more effective task: instead of learning the whole attention weights, the ARN learns an attention weight correction amount to rectify the original attention weights. Unlike the RobustScanner model [36], the attention rectification module with a parallel attention mechanism in the ARN needs only to learn to align features with characters rather than predict the text.

The ARN follows the encoder-decoder structure, but an attention rectification module is additionally introduced into the model. The attention rectification module contains an LSTM encoder, an attentional layer, a positional encoding block and a character spatial constraint block. The attention weight correction amount generated by the attention rectification module is defined as the unnormalized attention weights learned by the attentional layer. As shown in Fig. 4.1c and d, the attention weight correction amount is incorporated into the original attention weights to guide a more accurate decoding process.

The character spatial constraint block incorporates the CTC objective function [6] in the training stage. CTC learns to map each element in feature sequences output by the encoder into a character or a "blank" symbol. Therefore, CTC can be applied to enhance character position information in features.

Finally, this chapter integrates dynamic temporal residual learning and attention rectification, and constructs an ARN-DTRN model. The experimental results on the IAM English handwriting dataset [18] and the Rimes French handwriting dataset [9] demonstrate the effectiveness of the method proposed in this chapter.

The main contributions of this chapter are as follows:

1. This chapter improves the ability of encoder-decoder models to model long-term contextual dependencies. For the encoding method, a temporal residual learning method is proposed for an RNN encoder, which introduces weighted residual

4.2 Temporal Residual Learning

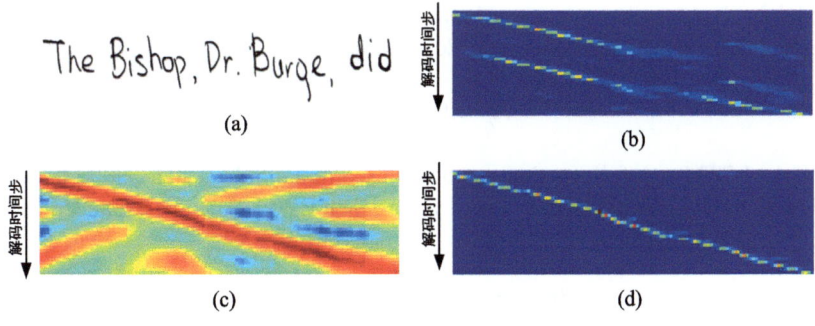

Before rectification: the Bishop . The Bishop . Dr__uge , did
After rectification: The Bishop . Dr . Burge , did

(e)

Fig. 4.1 Example of the attention drift problem and the attention rectification process. (**a**) Input image. (**b**) Original attention weights. (**c**) Attention patch. (**d**) Rectified attention weights. (**e**) Predictions before and after attention rectification

connections into the temporal dimension of the RNN. A secondary network is applied to learn dynamic weights for residual connections.
2. For the decoding method, an attention rectification method is proposed, which enhances position information in features to rectify attention weights learned by the original decoder.
3. The experimental results show that dynamic temporal residual learning can improve the performance of LSTM networks and that the attention rectification method can alleviate the attention drift problem.

This chapter first introduces the dynamic temporal residual learning method, then presents the attention rectification method, and finally, presents the experimental results. The main content of this chapter comes from the authors' published papers [32, 33].

4.2 Temporal Residual Learning

Considering $\mathcal{H}(x_t) = \text{rnn}(x_t)$ as an underlying mapping to be fit by an LSTM unit, where $x_t \in \mathbb{R}^d$ is the current input at time step t, the unit approximates a residual mapping explicitly: $\mathcal{F}(x_t) = \mathcal{H}(x_t) - \mathcal{H}(x_{t-1})$, where $\mathcal{H}(x_{t-1})$ is the hidden state output by the LSTM unit at time step $t - 1$. For brevity, we can abbreviate $\mathcal{F}_t = \mathcal{F}(x_t)$ and $\mathcal{H}_t = \mathcal{H}(x_t)$. The original mapping function can be formulated as

$$\mathcal{H}_t = \mathcal{F}_t + \mathcal{H}_{t-1} \tag{4.1}$$

Temporal residual learning can be implemented by adding a shortcut connection between the previous output and the current output of an LSTM unit. To better model the dynamic dependency, a dynamic weight α_t can be further introduced into the shortcut connection. In this way, Eq. (4.1) becomes

$$\mathcal{H}_t = \mathcal{F}_t + \alpha_t \mathcal{H}_{t-1} \tag{4.2}$$

If the LSTM unit follows the commonly used initial hidden state $\mathcal{H}_0 = \mathbf{0}$, i.e., if the hidden state is initialized as a zero vector, then Eq. (4.2) can be expanded as

$$\mathcal{H}_t = \mathcal{F}_t + \sum_{k=1}^{t-1} \left(\prod_{i=k+1}^{t} \alpha_i \right) \mathcal{F}_k \tag{4.3}$$

Equation (4.3) indicates that the desired underlying mapping of the current time step t can be decomposed into two additive parts. The first part, $\mathcal{F}_t$, is the nonlinear representation newly learned from the current input, whereas the second part contains the additional contextual information of the previous time steps.

In the second part of Eq. (4.3), $\mathcal{F}_k$ is weighted by $\prod_{i=k+1}^{t} \alpha_i$. If $\alpha_t \equiv 0$, the result is the same as that of the classical LSTM. When $\alpha_t \equiv 1$, identity mapping is constructed. When $0 < \alpha_t < 1$, the closer time step contributes more to $\mathcal{H}_t$, whereas the more distant time steps rarely affect the current time step because of continuous multiplication. This property complies with the intuitive observation for most sequential data in handwriting and speech recognition tasks. Thus, by adding the dynamic weight α_t, the model can better capture complex dependencies by exploring the additional contextual information in the sequential data.

The computational graph of a modified LSTM unit with the temporal residual learning mechanism is shown in Fig. 4.2. The core of the modified LSTM unit is the shortcut connection between the temporally adjacent outputs.

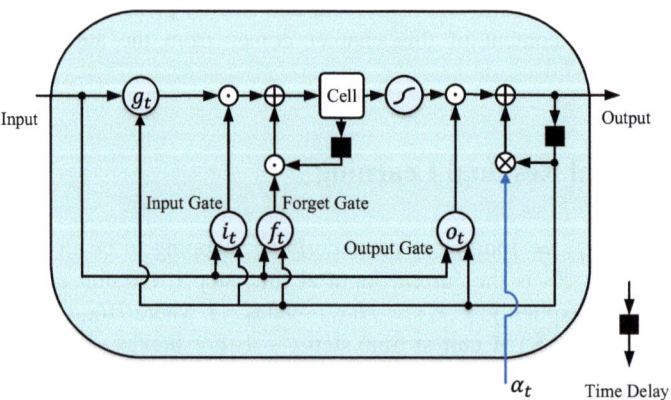

Fig. 4.2 Computational graph of a modified LSTM unit with a temporal residual learning mechanism

4.3 Dynamic Weights for Temporal Residual Connections

Let x_t and h_t be the input and output vectors of the LSTM unit at the t time step; then, the following recurrent transition functions describe the formula for a classical LSTM unit [13]:

$$g_t = \phi\left(W_{xg}x_t + W_{hg}h_{t-1} + b_g\right) \quad (4.4)$$

$$i_t = \sigma\left(W_{xi}x_t + W_{hi}h_{t-1} + b_i\right) \quad (4.5)$$

$$f_t = \sigma\left(W_{xf}x_t + W_{hf}h_{t-1} + b_f\right) \quad (4.6)$$

$$o_t = \sigma\left(W_{xo}x_t + W_{ho}h_{t-1} + b_o\right) \quad (4.7)$$

$$c_t = i_t \odot g_t + f_t \odot c_{t-1} \quad (4.8)$$

$$h_t = o_t \odot \phi(c_t) \quad (4.9)$$

where $\phi(\cdot)$ and $\sigma(\cdot)$ represent the hyperbolic tangent (tanh) function and the sigmoid function, respectively. $\odot$ denotes the elementwise product.

With the temporal residual connections, the current output h_t is computed by adding the weighted delayed output h_{t-1} to the output of a classical LSTM unit; thus, Eq. (4.9) is modified into the following equation:

$$h_t = o_t \odot \phi(c_t) + \alpha_t h_{t-1} \quad (4.10)$$

where $o_t \odot \phi(c_t)$ is the nonlinear representation that the network should learn at time step t. The weight of the shortcut connection at time step t is denoted by α_t.

4.3 Dynamic Weights for Temporal Residual Connections

There are two ways to obtain α_t, i.e., statically or dynamically. A straightforward way is to set α_t as a constant, i.e., $\alpha_t \equiv \gamma$ ($0 < \gamma \leq 1$). Under this condition, the network with fixed residual connection weights is called a static temporal residual network (STRN).

To better model the dynamic dependencies of the input sequence, this chapter further proposes the use of dynamic weights for residual connections. A secondary network is applied to generate α_t dynamically from the input sequence. Compared with gated units in the classical LSTM network, the use of an independent secondary network is more flexible for learning contextual dependencies, and different network structures can be designed for various sequence modeling tasks. At each time step, an internal vector representation z_t is first learned by a network from the input feature x_t and then converted to a scalar α_t by a fully connected layer with one neuron. The scalar α_t is used as the weight of the shortcut connection for the corresponding time step in the primary network. In general, the dynamic weights are generated as follows:

$$z_t = g(x_t) \tag{4.11}$$

$$\alpha_t = \sigma\left(w^T z_t + b\right) \tag{4.12}$$

where $g(\cdot)$ is the composite mapping function used to learn the internal vector representation z_t from the input x_t. w and b are the trainable parameters of the fully connected layer. A sigmoid activation function $\sigma(\cdot)$ is used to normalize the dynamic weight α_t between 0 and 1.

Two structures for the secondary network are investigated, i.e., an LSTM network and a self-attention network.

A discount factor is used to constrain the maximum weights of the shortcut connections. As a result, Eq. (4.10) actually becomes

$$h_t = o_t \odot \phi(c_t) + \gamma \cdot \alpha_t h_{t-1} \tag{4.13}$$

where γ $(0 < \gamma \leq 1)$ is the preset discount factor.

A dynamic temporal residual network (DTRN) consists of a primary network and secondary network. The primary network is a modified LSTM network that introduces a dynamic temporal residual learning mechanism, and the secondary network is used to generate dynamic weights for shortcut connections in the primary network. Compared with the primary network, the secondary network is controlled to be lightweight to avoid a large additional computational complexity. Figure 4.3 shows a DTRN with a 2-layer primary network and a 1-layer secondary network. Both the primary network and the secondary network are bidirectional. For each direction in the secondary network, there is a corresponding LSTM or self-attention network. The obtained $\overrightarrow{\alpha_t}$ and $\overleftarrow{\alpha_t}$ are used for the corresponding directions of the primary network.

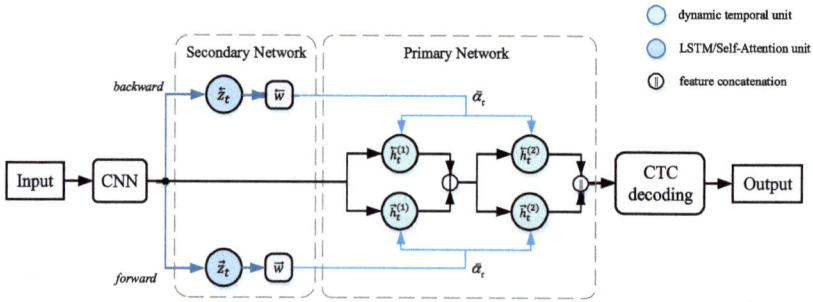

Fig. 4.3 The system framework of a bidirectional DTRN with a 2-layer primary network

A feature extraction module can be added into DTRN. A CNN is adopted as the feature extraction module for handwriting recognition tasks. In experiments, the performance of DTRN was also verified in a speech recognition task, and the Mel frequency cepstral coefficient (MFCC) [3] is adopted as the feature extraction module.

To explore the effects of encoding methods, DTRN uses CTC [6] for decoding. During training, the primary and secondary networks are jointly trained through a common objective function, which is calculated as

$$\mathcal{L} = - \sum_{(I,y) \in S} \log p(y|I) \tag{4.14}$$

where I is the input image, y is the ground-truth text, and S represents the training set.

4.4 Attention Rectification

This chapter further studies the decoding methods for encoder-decoder models. As discussed before, although the attentional decoder can exploit the semantic information in the text, the attention drift problem occurs frequently for images with long texts. To solve this problem, an attention rectification method is proposed, which introduces an additional attention rectification module into the model. The attention rectification module aims to learn an attention weight correction amount to rectify the attention weights generated by the original decoder. A positional encoding block with a GRU network and a character spatial constraint block with the CTC criterion are used to enhance the character position information in feature sequences.

An attention rectification network (ARN) is established on the basis of the attention rectification method, as shown in Fig. 4.4. The ARN consists of a CNN feature extraction module, a regular encoder-decoder module, and an attention rectification module. For the regular encoder-decoder module, an LSTM network [13] is used for encoding, and a Transformer decoder [29] is used for decoding.

Unlike the original Transformer decoder, the calculation of attention weights for the decoder in the regular encoder-decoder module is divided into two parts. The first part is the regular attention weights computed by the multihead attention mechanism, and the second part is the attention weight correction amount learned by the attention rectification module.

The attention rectification module consists of an LSTM encoder, a positional encoding block, a character spatial constraint block, and an attentional layer. The attention weight correction amount is defined as the unnormalized attention weights generated by the attentional layer, which is calculated by the dot-product of queries and keys. The queries are positional encodings generated by using character ordinal numbers through a GRU network, and the keys are outputs of the character spatial

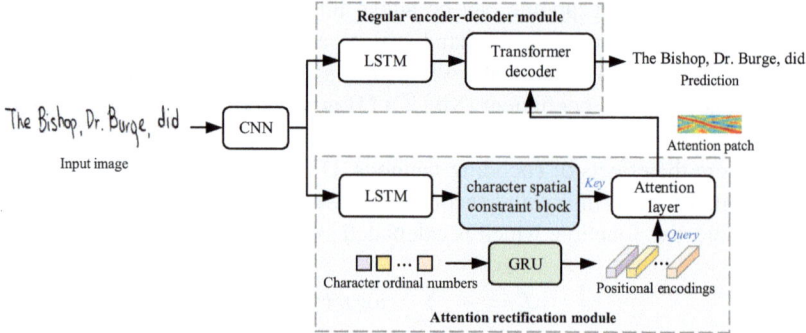

Fig. 4.4 The system framework of an attention rectification network

constraint block. The character spatial constraint block is composed of three fully connected layers, and the CTC objective function is introduced in the training stage.

4.4.1 Positional Encoding

One of the reasons for the attention drift problem is that the calculation of attention weights is sensitive to the previously decoded results [30, 36]. To reduce the impact of the previously decoded results on the calculation of attention weights, the attention rectification module uses only positional encodings or characters as the input of the attentional layer.

A GRU network [2] is used to generate positional encodings from character ordinal numbers. The positional encoding corresponding to the t-th character $\boldsymbol{p}_t$ can be calculated via the following equations.

$$r_t = \sigma \left(W_{ir} \boldsymbol{x}_t + W_{pr} \boldsymbol{p}_{t-1} + \boldsymbol{b}_r \right) \tag{4.15}$$

$$z_t = \sigma \left(W_{iz} \boldsymbol{x}_t + W_{pz} \boldsymbol{p}_{t-1} + \boldsymbol{b}_z \right) \tag{4.16}$$

$$\tilde{\boldsymbol{p}}_t = \phi \left(W_{ip} \boldsymbol{x}_t + W_{pp} \left(r_t \odot \boldsymbol{p}_{t-1} \right) + \boldsymbol{b}_p \right) \tag{4.17}$$

$$\boldsymbol{p}_t = (1 - z_t) \odot \tilde{\boldsymbol{p}}_t + z_t \odot \boldsymbol{p}_{t-1} \tag{4.18}$$

where $\boldsymbol{x}_t$ is a one-hot vector in which the t-th element is 1 and the remaining elements are all 0. W_* and $\boldsymbol{b}_*$ are trainable parameters. $\phi(\cdot)$ is the tanh activation function, and $\sigma(\cdot)$ is the sigmoid function. $\odot$ refers to the elementwise product.

Using a GRU network to generate positional encoding causes the calculation of the current positional encoding $\boldsymbol{p}_t$ to use the previous output $\boldsymbol{p}_{t-1}$, which alleviates the problem that positional encodings corresponding to large ordinal numbers are underfitting because of insufficient training samples. It can also help learn temporal dependencies of the input sequence since each positional encoding is not independent.

All positional encodings are concatenated in the temporal dimension to obtain $P = \{p_t\}_{t=1}^{L}$, where L is the preset maximum decoding length.

Although the GRU network runs recursively during the training stage, all positional encodings can be generated in advance during the test stage. Therefore, the attention rectification module can run in parallel.

4.4.2 Character Spatial Constraints

Abundant character position information helps learn more accurate alignments between the feature sequence and text. The CTC [6] criterion can map each element in the feature sequence to a character or a "blank" symbol. Therefore, incorporating CTC in the model helps enhance the character position information in features.

On the basis of the above analysis, a simple yet effective spatial constraint block is proposed. In the attention rectification module, the hidden representations output by the LSTM encoder are $H \in \mathbb{R}^{T \times d}$, where T is the length of the hidden representations and d is the dimension of the features. The calculation process of the character spatial constraint block is as follows:

$$G = HW_g + b_g \tag{4.19}$$

$$K = \phi(GW_{k1} + b_{k1})W_{k2} + b_{k2} \tag{4.20}$$

where $W_g \in \mathbb{R}^{d \times C}$ converts the hidden representations H into $G \in \mathbb{R}^{T \times C}$, in which C is the number of classes. $G \in \mathbb{R}^{T \times C}$ is the unnormalized character probability. Then, a two-layer feed-forward network transforms G into a feature sequence $K \in \mathbb{R}^{T \times d}$, and K is used as the key for the subsequent attentional layer. $\phi(\cdot)$ is the Swish activation function [25]. The character spatial constraint block contains only three fully connected layers, so it does not increase the computational complexity of the model.

In the training stage, a CTC objective function is introduced into the ARN, which acts on the unnormalized character probabilities G. In this way, the character position information in the keys of the attentional layer K can be enhanced.

4.4.3 Attention Rectification Process

In the attention rectification module, the values of positional encoding P are used as queries, outputs of the character spatial constraint block K are used as keys, and the attention weight correction amount $\tilde{\alpha}$ can be calculated via the following equation.

$$\tilde{\alpha} = PK^T \tag{4.21}$$

Let H_r be the hidden representations output by the encoder in the regular encoder-decoder module. The Transformer decoder in the regular encoder-decoder module consists of a self-attention layer, a cross-attention layer, and a feedforward layer. The self-attention layer learns semantic dependencies of text from character embedding vectors of the previously decoded results. Let E be the output of the self-attention layer. The calculation process of the cross-attention layer can be formulated as

$$\alpha = \mathrm{softmax}\left(\frac{1}{2\sqrt{d}}\left(EW_Q W_K^T H_r^T + \tilde{\alpha}\right)\right) \qquad (4.22)$$

where W_Q and W_K are learnable parameters. α refers to the rectified attention weights. The regular encoder-decoder module uses α to weight the encoder outputs H_r, and the results of the weighted summation of H_r are transformed into decoder outputs by the feedforward layer.

4.4.4 Training and Inference

The regular encoder-decoder module and the attention rectification module in the ARN can be jointly trained. In the training stage, two auxiliary objective functions are applied in addition to the cross-entropy between the ground-truth text and the character probabilities predicted by the regular encoder-decoder module. The first auxiliary objective function is the CTC loss introduced in the character spatial constraint block. The second auxiliary objective function is the cross-entropy between the ground-truth text and the character probabilities predicted from the attention weight correction amount, which is calculated as

$$V = \mathrm{softmax}\left(\frac{1}{\sqrt{d}}\tilde{\alpha}\right) H \qquad (4.23)$$

$$Y = V W_o + b_o \qquad (4.24)$$

where H refers to the hidden representations output by the encoder in the attention rectification module. $\tilde{\alpha}$ is the attention weight correction amount. Y denotes the unnormalized character probabilities predicted from $\tilde{\alpha}$. W_o and b_o are learnable parameters. Notably, the above decoding process involving the attention weight correction amount $\tilde{\alpha}$ is only used to learn a more accurate alignment in the training stage. In the test stage, the attention rectification module needs to calculate only the attention weight correction amount $\tilde{\alpha}$ rather than the unnormalized character probabilities Y.

The overall objective function of the ARN in the training stage is

$$\mathcal{L} = \mathcal{L}_r + \lambda_1 \mathcal{L}_a + \lambda_2 \mathcal{L}_{\mathrm{CTC}} \qquad (4.25)$$

where $\mathcal{L}_r$ is the cross-entropy between the ground-truth text and the character probabilities predicted by the regular encoder-decoder module, $\mathcal{L}_a$ is the cross-entropy between the ground-truth text and the character probabilities predicted by the attention weight correction amount, and $\mathcal{L}_{CTC}$ is the CTC loss in the character spatial constraint block. λ_1 and λ_2 are preset constants. In the subsequent experiments, λ_1 and λ_2 are both set to 0.2. In the test stage, the ARN uses the decoded results from the regular encoder-decoder module as predictions.

Furthermore, dynamic temporal residual learning and attention rectification methods can be integrated, i.e., the dynamic temporal residual learning mechanism can be incorporated into the LSTM encoder in the regular encoder-decoder module of the ARN. The model that combines dynamic temporal residual learning and attention rectification is named ARN-DTRN.

4.5 Discussion

This section discusses the effects of dynamic temporal residual learning and attention rectification methods. The performance indices of DTRN and ARN are evaluated on handwriting recognition datasets of different languages. Finally, the effects of the ARN-DTRN model with both dynamic temporal residual learning and attention rectification are discussed.

4.5.1 Dynamic Temporal Residual Learning

The performance of DTRN is evaluated on the offline handwriting recognition task with three popular public datasets: the IFN/ENIT Arabic handwriting dataset [22], the IAM English handwriting dataset [18] and the Rimes French handwriting dataset [9]. To verify the effectiveness of DTRN on other sequence modeling tasks, an additional speech recognition experiment is also conducted on the TIMIT dataset [5]. The TIMIT dataset contains 3696 training samples and 192 test samples.

For handwriting recognition, the character error rate (CER) is used as the evaluation metric. For speech recognition, the phoneme error rate (PER) is used as the evaluation metric. PER can be calculated by the edit distance between the prediction and the label at the phoneme-level.

Table 4.2 Configuration of the CNN for Arabic handwriting recognition

Parameter	Layer 1	Layer 2	Layer 3	Layer 4	Layer 5
#kernels	16	32	48	64	80
MaxPool	Yes	Yes	No	No	No
Dropout	0	0	0.2	0.2	0.2
Kernel size/stride	$3 \times 3 / 1 \times 1$				
Pooling size/stride	$2 \times 2 / 2 \times 2$				

4.5.1.1 Arabic Handwriting Recognition

All the input images are preprocessed via the center-normalizer method provided by the OCRopus system[1] [34] and resized to a unified height of 48 pixels while preserving the original aspect ratio.

The feature extraction module in the model uses a 5-layer CNN, and its configuration is listed in Table 4.2. Batch normalization [14] is used for each convolution layer. The activation function after batch normalization is leaky ReLU.

For the baseline LSTM, the STRN and the primary network in the DTRN, the first and second RNN layers have 64 units, and the other RNN layers have 128 units. For example, the numbers of units in a 5-layer LSTM are 64, 64, 128, 128 and 128. Dropout is used for each layer with a fixed dropout rate of 0.5.

All the models are trained for 300 epochs with a learning rate of 1×10^{-4}. The batch size is 16, and the optimizer is RMSprop [26].

Comparison of the Static Time Residual Network (STRN) and Dynamic Time Residual Network (DTRN) This experiment compares the performance of the STRN with fixed weights and the DTRN with dynamic weights. The residual connection weight of the STRN is fixed as $\alpha_t \equiv \gamma$. For DTRN, the maximum value of α_t is also constrained by the discount factor γ. The numbers of RNN layers in the STRN and the primary network of the DTRN are set to five. For DTRN, the secondary network used to learn dynamic weights is an LSTM network, which has one layer with 32 units.

Figure 4.5 shows the comparison results on the test set. The vertical line on the top of each bar indicates the standard deviation from three independent experiments. For the STRN, large residual connection weights degrade the model's performance. DTRN is more stable to variations in the discount factor, and it achieves slightly higher accuracy than the best result of the STRN. On the basis of the results in Fig. 4.5, the fixed weight for the STRN is set to 0.3, and the discount factor for the DTRN is set to 0.4 in the subsequent experiments.

Different Configurations of the Secondary Network For a DTRN, the secondary network performs a simpler task than does the primary network; thus, its numbers

---
[1] https://github.com/tmbdev/ocropy

4.5 Discussion

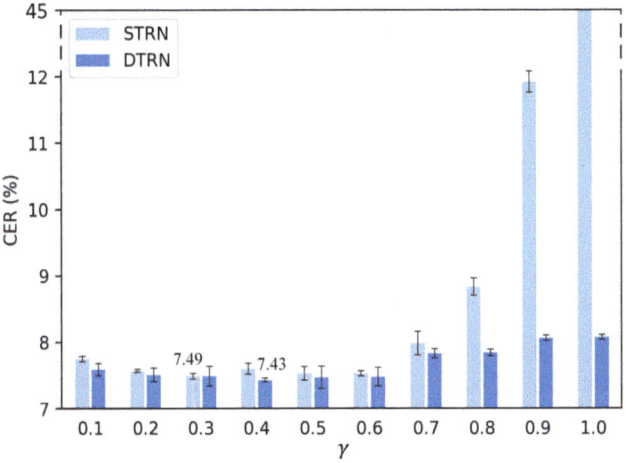

Fig. 4.5 Comparison of recognition performance for the STRN and DTRN

Table 4.3 Comparison of different configurations for the secondary LSTM network

#Layers	Hidden size	#Params	Train	Test
1	8	1.87 M	0.82	7.52
	16	1.93 M	0.81	**7.41**
	32	2.06 M	0.79	7.44
	64	2.33 M	0.81	**7.41**
2	32, 32	2.08 M	0.80	7.49
	32, 64	2.12 M	0.77	7.42
	64, 32	2.37 M	0.80	7.57

Bold values highlight the optimal performance (e.g., highest accuracy or lowest error rate) for the respective metric across different methods, models, or systems

of layers and units can be set to smaller values than those in the primary network. This experiment compares different configurations for the two types of secondary networks, i.e., an LSTM network or a self-attention network. The number of RNN layers in the primary network is fixed at 5. For the secondary LSTM network, different numbers of RNN layers and units are compared, and the results are listed in Table 4.3. For the secondary self-attention network, the performance indices of the models with different hidden sizes are listed in Table 4.4. Since the self-attention mechanism has high computational complexity, the secondary self-attention network is fixed to one layer.

The comparison results indicate that an appropriate size for the secondary network may achieve slightly better performance on the test set. Using an LSTM as the secondary network yields a slightly lower CER on the test set for Arabic handwriting recognition.

Table 4.4 Comparison of different configurations for the secondary self-attention network

Hidden size	#Params	Train	Test
64	1.99 M	0.83	7.50
128	2.17 M	0.83	7.47
256	2.54 M	0.82	**7.44**

Bold values highlight the optimal performance (e.g., highest accuracy or lowest error rate) for the respective metric across different methods, models, or systems

Table 4.5 Comparison of baseline LSTM networks and DTRNs with different numbers of layers on the IFN/ENIT dataset

Model	#Layers	#Params	Train	Test
LSTM	4	1.41 M	1.35	8.29
	5	1.80 M	1.04	7.84
	6	2.20 M	0.99	7.71
	7	2.59 M	0.92	7.52
	8	2.99 M	0.97	7.74
DTRN	4	1.66 M	1.05	7.74
	5	2.06 M	0.79	7.43
	6	2.45 M	0.75	7.39
	7	2.85 M	0.72	**6.91**
	8	3.24 M	0.72	7.53

Bold values highlight the optimal performance (e.g., highest accuracy or lowest error rate) for the respective metric across different methods, models, or systems

Comparison of Different Numbers of Layers in the Primary Network This experiment compares the performance of the baseline LSTM and DTRN with different numbers of layers in the primary network. To explore the performance of models with a controlled number of parameters, a 1-layer LSTM with 32 units is adopted as the secondary network of the DTRN. In this way, the DTRN has slightly fewer parameters than the baseline LSTM network with one more layer.

The comparison results are listed in Table 4.5. Compared with the baseline LSTM networks, DTRNs with similar (or even smaller) numbers of network parameters achieve better results, which indicates that the introduction of the temporal residual learning mechanism can improve performance.

The 7-layer DTRN with a training CER of 0.72% and a test CER of 6.91% achieves the best result, showing that DTRN has better nonlinear sequence modeling and generalization ability.

4.5.1.2 English and French Handwriting Recognition

This chapter further verifies the performance of DTRN on the IAM English handwriting dataset [18] and the Rimes French handwriting dataset [9]. Compared

4.5 Discussion

with the handwritten Arabic text images in the IFN/ENIT dataset [22], the IAM and Rimes datasets contain considerable images with long texts, which require the model to better capture long-term temporal dependencies.

All the images are resized to a fixed height of 128 pixels while preserving the original aspect ratio. The learning rate is set to 3×10^{-4}, and the optimizer is RMSprop [26]. The batch size is set to 16. The training process continues until the CER on the validation set stops decreasing. In general, a model takes approximately 300 epochs to train.

Following other previous work [24] implemented in Torch, a CNN-LSTM-CTC model is reimplemented in this experiment via PyTorch [21]. The feature extractor is a 5-layer CNN. Compared with the CNN used in Arabic handwriting experiments, the only difference is that the max pooling operation is adopted for the first three layers instead of the first two layers.

For each LSTM layer, the number of units is 256, and the dropout rate is set to 0.5. The STRN and the primary network in DTRN have the same network configuration as the corresponding part of the baseline LSTM. A 1-layer LSTM with 128 units is adopted as the secondary network of the DTRN. In this way, the number of parameters in a DTRN is slightly less than that in a baseline LSTM with one more layer.

The performances of the LSTM and DTRN models with different numbers of layers are compared, and the results are listed in Table 4.6. The 5-layer DTRN achieves the best performance on the IAM validation set, Rimes validation set, and Rimes test set. The 6-layer DTRN achieves the best performance on the IAM test set. DTRN with l layers in the primary network has slightly fewer parameters than the LSTM network with $l + 1$ layers while achieving lower CERs. For example, the 5-layer DTRN has 0.13 M fewer parameters than the 6-layer LSTM network does, but its CER is 0.34% lower on the IAM validation set, 0.4% lower on the IAM test set, 0.29% lower on the Rimes validation set, and 0.27% lower on the Rimes test set. The results in Table 4.5 demonstrate that dynamic temporal residual learning can improve the performance of LSTM networks.

Table 4.6 Comparison of baseline LSTM networks and DTRNs with different numbers of layers on the IAM and Rimes datasets

Model			IAM		Rimes	
RNN	#Layers	#Params	Val	Test	Val	Test
LSTM	3	6.44 M	5.31	8.78	3.44	3.36
	4	8.01 M	4.95	8.07	3.05	2.99
	5	9.59 M	4.58	7.62	2.75	2.63
	6	11.17 M	4.50	7.35	2.65	2.55
DTRN	3	7.88 M	4.73	7.75	2.79	2.77
	4	9.46 M	4.42	7.29	2.50	2.37
	5	11.04 M	**4.16**	6.95	**2.36**	**2.28**
	6	12.61 M	4.24	**6.91**	2.43	2.30

Bold values highlight the optimal performance (e.g., highest accuracy or lowest error rate) for the respective metric across different methods, models, or systems

Table 4.7 Speech recognition results on the TIMIT dataset

Model	#Params	Train	Test
LSTM [7]	4.3 M	–	17.70
baseline LSTM	3.62 M	28.05	15.22
STRN	3.62 M	24.42	14.85
DTRN (self-attention)	3.80 M	23.97	**14.59**
DTRN (LSTM)	3.67 M	23.72	14.74

Bold values highlight the optimal performance (e.g., highest accuracy or lowest error rate) for the respective metric across different methods, models, or systems

4.5.1.3 Speech Recognition

The dynamic temporal residual learning mechanism is a general sequence modeling method. To verify the performance of DTRN in other sequence modeling tasks, an additional speech recognition experiment is conducted on the TIMIT dataset [5].

The baseline model is a 3-layer LSTM network slightly modified from the model used in other previous work [7]. Each LSTM layer has 256 units (while each LSTM layer has 250 units in [7]). The STRN and the primary network in DTRN have the same configuration as the baseline LSTM network. For the secondary network in DTRN, the hidden size is set to 64 for both the LSTM secondary network and the self-attention secondary network.

The dropout rates for all the layers are set to 0.5. All the models are trained for 150 epochs with a learning rate of 1×10^{-4}, and the optimizer is RMSprop [26].

The input feature is a conventional MFCC with a dimension of 39. All 61 phoneme labels are used during training and decoding and are mapped to 39 classes following the methods of other previous works [15] to obtain the test phoneme error rate.

As shown in Table 4.7, the STRN and DTRN outperform the classical LSTM network. DTRN with a self-attention-based dynamic weight generator achieves the best performance on the test set, with a CER of 14.59%, which is different from the results of handwriting recognition experiments. It seems that the self-attention method has better sequence modeling ability for one-dimensional speech signals. This result indicates that designing appropriate secondary network structures for different tasks can be beneficial to the performance of DTRN.

4.5.1.4 Visualization

The Dynamic Weights for Temporal Shortcuts Figure 4.6 shows the values of α_t in a bidirectional DTRN with respect to a sample from the IAM dataset. Larger values of the dynamic coefficient α_t indicate that more prior outputs are passed directly to the outputs of the current time step. α_t usually obtains its maximum value in uniformly structured areas (e.g., empty space) and obtains its minimum value while encountering a new character, which complies with intuitive observations for most sequential data.

4.5 Discussion

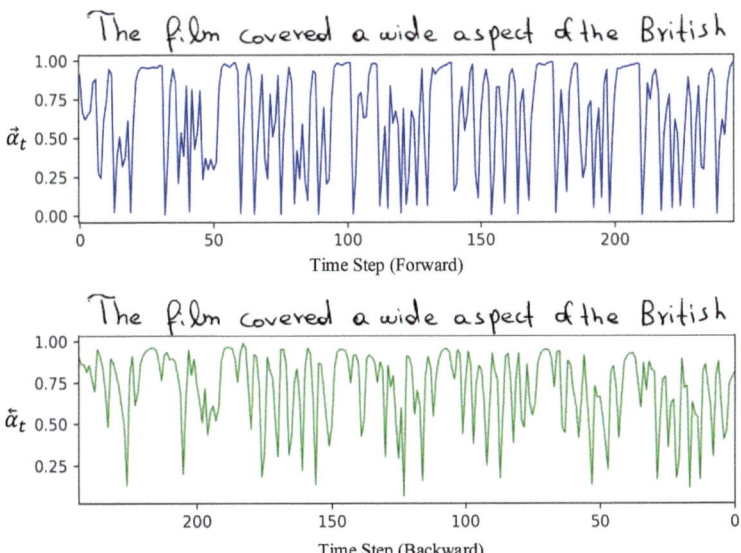

Fig. 4.6 Examples of α_t in DTRN for both the forward and backward directions

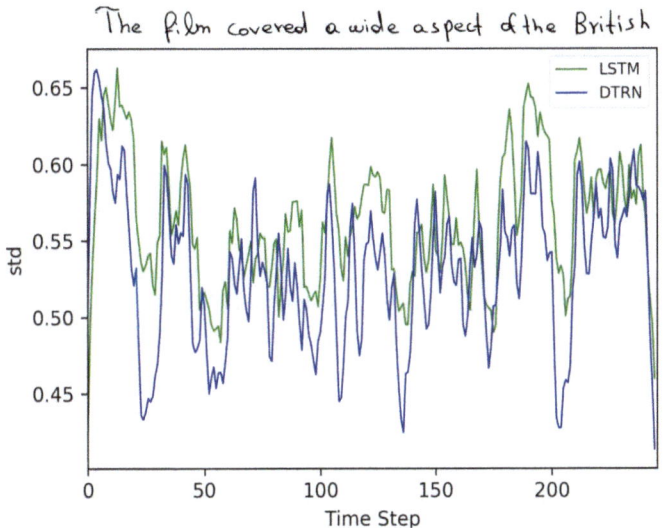

Fig. 4.7 Standard deviations (stds) of the nonlinear responses of the baseline LSTM and DTRN for the same input image

The Statistics of the Modified LSTM Unit Outputs In ResNet, the learned residual functions generally have small responses [11], and a similar phenomenon can be found in DTRN. Figure 4.7 shows the standard deviation of the learned nonlinear responses ($\mathcal{F}_t$ in the $formula$ (4.3)) at each time step of both a 5-layer

Table 4.8 Examples of the three error types on three handwriting datasets

Database	Data	Incorrect characters	Missing characters	Extra characters
IAM	Input Image	*compartments* (handwritten)	*understand* (handwritten)	*careful* (handwritten)
	Ground Truth	compartments	understand	careful
	Prediction	compostments	under_tand	careferl
Rimes	Input Image	*entraine* (handwritten)	*rembourser* (handwritten)	*ouverture* (handwritten)
	Ground Truth	entraine	rembourser	ouverture
	Prediction	entraîne	rembourer	ouvertiure
IFN/ENIT	Input Image	تالة (handwritten)	الشريفات (handwritten)	الزريبة (handwritten)
	Ground Truth	تالة	الشريفات	الزريبة
	Prediction	شالة	الشريات	الزريببة

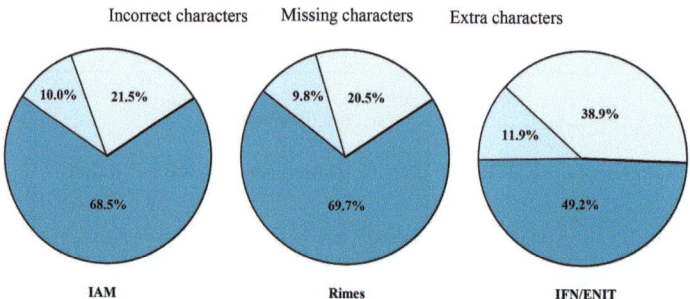

Fig. 4.8 The proportions of the three error types on the test set of three handwriting datasets using DTRNs

baseline LSTM and 5-layer DTRN with respect to the same input image. At most time steps, the nonlinear responses that DTRN outputs are smaller than those of the baseline LSTM network, which indicates that DTRN can learn more concise representations. This phenomenon might explain why DTRNs with similar or fewer parameters achieve better performance than do baseline LSTM networks.

Error Analysis When the edit distance is used to compute the CER for offline handwriting recognition, three types of errors may occur: incorrect characters, missing characters, and extra characters. Table 4.8 shows some examples for each case, where wrong characters and extra characters are marked in red and "_" indicates missing characters. The samples in the IAM and Rimes datasets usually contain text lines with multiple words, and only the words with errors are shown in Table 4.8.

The proportion of each error type is shown in Fig. 4.8. Incorrect characters are the dominant factor, and most of them are caused by similar characters, such as (a, o), (i, î), etc. Missing character errors generally occur when neighboring characters are touching or cursive, both of which are common in Arabic handwriting samples.

Extra character errors occur less frequently and usually appear along with incorrect character errors.

4.5.2 Attention Rectification

For the attention rectification network (ARN), different positional encoding methods in the attention rectification module are compared first. Then, different decoding methods, including CTC-based decoding, regular attention-based decoding, and the decoding method for the ARN, are compared. The evaluation metrics include the character error rate (CER) and the word error rate (WER).

4.5.2.1 Experimental Settings

For the feature extraction module in the ARN, three different backbone networks are adopted: ResNet-32 [1], ResNet-46 [30], and EfficientNet-B3 [28]. For the encoder, a 2-layer bidirectional LSTM network is used, and the number of units in each direction is $\frac{1}{2}$ of the number of channels in feature maps extracted by the CNN. The ARN uses a 2-layer Transformer decoder, and the maximum decoding length is set to 120.

All the models are trained for 200 epochs with a batch size of 16. The learning rate is initialized to 0.5 and decreases to 0.1 and 0.01 at the 120th epoch and 180th epoch, respectively. The optimizer is ADADELTA [37].

4.5.2.2 Comparison of Different Positional Encoding Methods

This experiment compares different positional encoding methods in the attention rectification module, including static positional encoding used in the original Transformer decoder (Static), positional encodings generated by a linear layer from the character ordinal numbers (Linear), and positional encodings generated by a GRU network from the character ordinal numbers (GRU). For "static + linear" and "static + GRU", the symbol "+" refers to elementwise addition. The feature extraction module for all the models in this experiment is ResNet-32.

The comparison results of the models with positional encoding methods on the IAM and Rimes datasets are listed in Table 4.9. Compared with static positional encodings using a linear layer, the use of a GRU to learn positional encodings achieves performance improvements. No performance improvement is observed when static positional encodings and the GRU are used jointly.

Table 4.9 Performance (%) of models with different positional encoding methods

Positional encoding	IAM				Rimes			
	Val		Test		Val		Test	
	CER	WER	CER	WER	CER	WER	CER	WER
Static	4.24	13.75	6.37	18.29	2.42	8.92	2.03	8.18
Linear	4.43	13.82	6.26	17.89	2.21	8.90	1.99	8.30
Static + Linear	4.30	13.75	6.36	18.10	2.23	8.71	1.93	8.32
GRU	**4.16**	**13.67**	**6.00**	**17.71**	**2.16**	8.72	**1.80**	7.80
Static + GRU	4.18	13.79	6.19	17.72	2.17	**8.70**	1.94	**7.77**

Bold values highlight the optimal performance (e.g., highest accuracy or lowest error rate) for the respective metric across different methods, models, or systems

Table 4.10 Performance (%) of models with different CNNs and decoding methods

CNN	Decoding method	IAM				Rimes			
		Val		Test		Val		Test	
		CER	WER	CER	WER	CER	WER	CER	WER
ResNet-32	CTC	4.34	15.61	6.03	20.04	2.47	11.45	1.98	10.02
	Attention	4.81	15.52	6.75	19.61	2.88	12.40	3.01	12.34
	ARN w/o Char	4.16	13.67	6.00	17.71	2.16	8.72	1.80	7.80
	ARN	4.22	13.71	5.95	17.57	2.00	**8.13**	1.78	7.54
ResNet-46	CTC	4.85	17.35	7.02	23.36	3.02	13.53	2.77	13.32
	Attention	5.68	18.72	7.66	23.11	2.84	11.61	2.63	11.40
	ARN w/o Char	4.64	14.92	7.02	20.35	2.37	9.74	2.06	8.74
	ARN	4.48	14.97	6.57	20.19	2.20	9.43	2.00	8.89
EfficientNet-B3	CTC	4.21	15.37	5.95	20.00	2.26	10.71	2.11	10.22
	Attention	4.78	15.01	6.50	18.67	2.57	10.66	2.33	10.27
	ARN w/o Char	4.20	13.43	5.85	17.04	2.10	8.63	1.99	8.21
	ARN	**3.86**	**12.76**	**5.37**	**16.40**	**1.91**	8.17	**1.69**	**7.36**

Bold values highlight the optimal performance (e.g., highest accuracy or lowest error rate) for the respective metric across different methods, models, or systems

4.5.2.3 Comparison of Different Decoding Methods

The comparison of models using different CNNs and decoding methods on the IAM and Rimes datasets are listed in Table 4.10, where "Attention" denotes encoder-decoder models based on the regular attention mechanism, and "ARN w/o Char" refers to ARN models without the character spatial constraint block. The following phenomena can be observed from Table 4.10.

1. Compared with CTC-based models, encoder-decoder models based on the regular attention mechanism can achieve similar or slightly lower WERs, but have significantly higher CERs. This result demonstrates that the regular attention mechanism is difficult to apply directly to handwriting recognition tasks.

4.5 Discussion

Table 4.11 Performance (%) of different models on the IAM English handwriting dataset and the Rimes French handwriting dataset

	IAM				Rimes			
	Val		Test		Val		Test	
Model	CER	WER	CER	WER	CER	WER	CER	WER
HMM/CNN [20]	4.40	13.90	6.30	17.20	–	–	–	–
MDLSTM [23]	7.40	27.30	10.80	35.10	5.90	25.40	6.80	28.50
1D-LSTM [24]	**3.80**	13.50	5.80	18.40	2.20	9.60	2.30	9.60
CRNN-FCNN [17]	3.86	13.51	6.14	20.04	3.90	12.17	3.34	11.23
2D-SACRN [16]	–	–	6.76	20.89	–	–	3.43	11.92
DAN [30]	–	–	6.40	19.60	–	–	2.70	8.90
ARN	3.86	12.76	5.37	16.40	1.91	8.17	1.69	7.36
ARN-DTRN	**3.80**	**12.57**	**5.30**	**16.17**	**1.70**	**7.58**	**1.68**	**7.22**

Bold values highlight the optimal performance (e.g., highest accuracy or lowest error rate) for the respective metric across different methods, models, or systems

2. After the attention rectification module is introduced, the model performance is significantly improved. ARNs without the character spatial constraint module can achieve higher performance than both CTC-based models and encoder-decoder models based on the regular attention mechanism.
3. The character spatial constraint block can further improve the model performance effectively.
4. The ARN using EfficientNet-B3 as the feature extraction module achieves the best performance on most of the validation and test sets.

4.5.2.4 Comparison with Other Methods

A comparison of the attention rectification network (ARN), ARN-DTRN with both dynamic temporal residual learning and attention rectification methods, and a series of recent models are listed in Table 4.11. For fair comparison, no additional synthetic samples or external corpora are used in the training stage. ARN-DTRN incorporates the dynamic temporal residual learning mechanism into the encoder of the ARN, which improves model performance on all the validation and test sets.

4.5.2.5 Visualization

Recognition examples of models with different decoding methods are shown in Fig. 4.9, where errors are marked in red and "_" indicates missing characters. The encoder-decoder model based on the regular attention mechanism suffers from the attention drift problem. For example, the word "could" is repeatedly recognized twice in the first sample, and the word "était" is missed in the second sample. Although the CTC-based model can alleviate the attention drift problem, it may

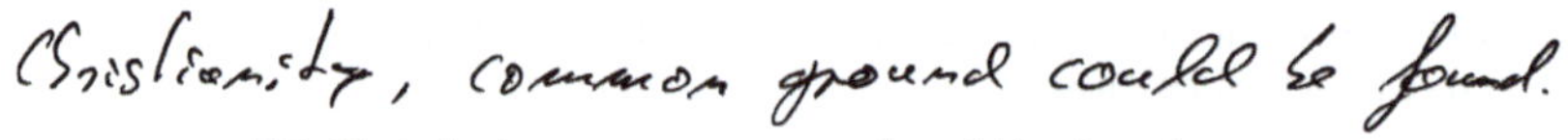

CSrisliamity, common ground could be found.

GT: Christianity , common ground could be found .

CTC: CSrislianity , common ground could be found .

Attention: G_ristianity , common ground could could be found .

ARN: Christianity , common ground could be found .

ouverture, il était évident qu'il y avait erreur dans le

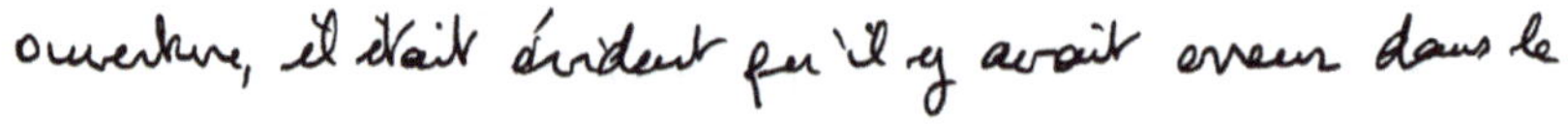

GT: ouverture, il était évident qu'il y avait erreur dans le

CTC: ouverture, il était écident qu'il y avait erreur dans le

Attention: ouverture, il ___ évident qu'il y avait erreur dans le

ARN: ouverture, il était évident qu'il y avait erreur dans le

la prendrait le 01/01/07. Je vous prie donc de bien voulor trouver

GT: la prendrait le 01/01/07. Je vous prie donc de bien vouloir trouver

CTC: la prendrait le 0_/01/07. Je vous prie donc de bien vouloir trouver

Attention: la prendrait le 01/___07. Je vous prie donc de bien touloir trouver

ARN: la prendrait le 01/01/07. Je vous prie donc de bien vouloir trouver

Fig. 4.9 Recognition examples of models with different decoding methods

confuse similar characters such as "C" and "G". Compared with the CTC-based model and the encoder-decoder model based on the regular attention mechanism, the ARN has more accurate prediction results.

For the three image samples in Fig. 4.9, the attention weights generated during the decoding process of the encoder-decoder model on the basis of the regular attention mechanism and ARN are shown in Fig. 4.10. The red arrows indicate the occurrence of attention drift. After the attention rectification method is introduced, the ARN can better learn the alignment between the feature sequence and the text.

4.6 Summary

This chapter studies both the encoding method and decoding method for the recognition of images with long texts.

For the encoding method, a dynamic temporal residual learning mechanism is proposed, which adds residual connections in the temporal dimension of the RNN.

4.6 Summary

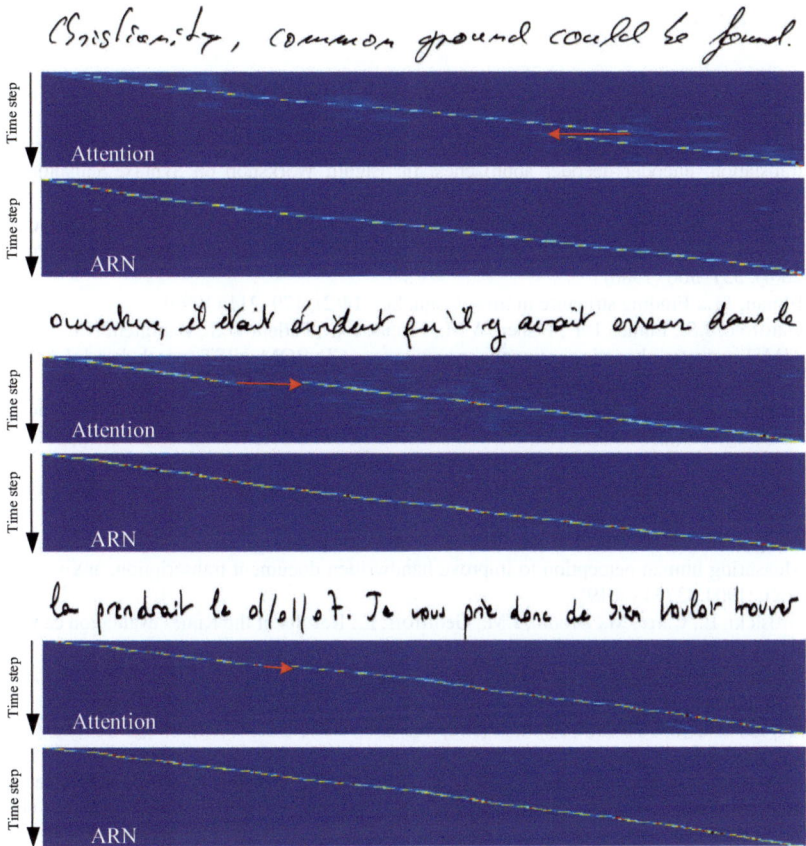

Fig. 4.10 Visualization of attention weights before and after the introduction of the attention rectification method

An independent secondary network is applied to generate dynamic weights of residual connections, and construct a dynamic temporal residual network (DTRN). The experimental results on handwriting and speech recognition datasets demonstrate that DTRN outperforms the classical LSTM network.

For the decoding method, an attention rectification method is proposed, which rectifies attention weights learned in the original decoder by introducing a parallel attention mechanism to enhance character position information in feature sequences.

By combining dynamic temporal residual learning with the attention rectification method, this chapter constructs an ARN-DTRN model, which further improves the model performance in handwriting recognition tasks.

References

1. Cheng, Z., Bai, F., Xu, Y., et al.: Focusing attention: towards accurate text recognition in natural images. In: ICCV, pp. 5076–5084 (2017)
2. Cho, K., van Merriënboer, B., Bahdanau, D., Bengio, Y.: On the properties of neural machine translation: encoder–decoder approaches. In: Eighth Workshop on Syntax, Semantics and Structure in Statistical Translation, pp. 103–111 (2014)
3. Davis, S., Mermelstein, P.: Comparison of parametric representations for monosyllabic word recognition in continuously spoken sentences. IEEE Trans. Acoust. Speech Signal Process. **28**(4), 357–366 (1980)
4. Elman, J.L.: Finding structure in time. Cogn. Sci. **14**(2), 179–211 (1990)
5. Garofolo, J.S., Lamel, L.F., Fisher, W.M., Fiscus, J.G., Pallett, D.S., Dahlgren, N.L.: DARPA TIMIT acoustic-phonetic continuous speech corpus CD-ROM. NIST speech disc 1-1.1. NASA STI/Recon Technical Report **93** (1993)
6. Graves, A., Fernández, S., Gomez, F., Schmidhuber, J.: Connectionist temporal classification: labelling unsegmented sequence data with recurrent neural networks. In: ICML, pp. 369–376 (2006)
7. Graves, A., Mohamed, A., Hinton, G.: Speech recognition with deep recurrent neural networks. In: ICASSP, pp. 6645–6649 (2013)
8. Grieggs, S., Shen, B., Rauch, G., Li, P., Ma, J., Chiang, D., Price, B., Scheirer, W.: Measuring human perception to improve handwritten document transcription. arXiv preprint arXiv:1904.03734 (2019)
9. Grosicki, E., Carré, M., Brodin, J.M., Geoffrois, E.: Results of the Rimes evaluation campaign for handwritten mail processing. In: ICDAR, pp. 941–945 (2009)
10. Gui, T., Zhang, Q., Zhao, L., Lin, Y., Peng, M., Gong, J., Huang, X.: Long short-term memory with dynamic skip connections. In: AAAI, pp. 6481–6488 (2019)
11. He, K., Zhang, X., Ren, S., Sun, J.: Deep residual learning for image recognition. In: CVPR, pp. 770–778 (2016)
12. Hochreiter, S.: The vanishing gradient problem during learning recurrent neural nets and problem solutions. Int. J. Uncertainty Fuzziness Knowledge-Based Syst. **6**(2), 107–116 (1998)
13. Hochreiter, S., Schmidhuber, J.: Long short-term memory. Neural Comput. **9**(8), 1735–1780 (1997)
14. Ioffe, S., Szegedy, C.: Batch normalization: accelerating deep network training by reducing internal covariate shift. In: ICML, pp. 448–456 (2015)
15. Lee, K.F., Hsiao wuen Hon: Speaker-independent phone recognition using hidden Markov models. IEEE Trans. Acoust. Speech Signal Process **37**(11), 1641–1648 (1989)
16. Ly, N.T., Nguyen, H.T., Nakagawa, M.: 2D self-attention convolutional recurrent network for offline handwritten text recognition. In: ICDAR, pp. 191–204 (2021)
17. Markou, K., Tsochatzidis, L., Zagoris, K., Papazoglou, A., Karagiannis, X., Symeonidis, S., Pratikakis, I.: A convolutional recurrent neural network for the handwritten text recognition of historical Greek manuscripts. In: ICPR, pp. 249–262 (2021)
18. Marti, U.V., Bunke, H.: The IAM-database: an English sentence database for offline handwriting recognition. Int. J. Doc. Anal. Recognit. **5**(1), 39–46 (2002)
19. Pascanu, R., Gulcehre, C., Cho, K., Bengio, Y.: How to construct deep recurrent neural networks. In: ICLR (2014)
20. Pastor-Pellicer, J., Castro-Bleda, M.J., Espana-Boquera, S., Zamora-Martinez, F.: Handwriting recognition by using deep learning to extract meaningful features. AI Commun. **32**(2), 101–112 (2019)
21. Paszke, A., Gross, S., Massa, F., et al.: PyTorch: an imperative style, high-performance deep learning library. In: NeurIPS, pp. 8024–8035 (2019)
22. Pechwitz, M., Maddouri, S.S., Märgner, V., Ellouze, N., Amiri, H., et al.: IFN/ENIT-database of handwritten Arabic words. In: CIFED, pp. 127–136 (2002)

23. Pham, V., Bluche, T., Kermorvant, C., Louradour, J.: Dropout improves recurrent neural networks for handwriting recognition. In: ICFHR, pp. 285–290 (2014)
24. Puigcerver, J.: Are multidimensional recurrent layers really necessary for handwritten text recognition? In: ICDAR, pp. 67–72 (2017)
25. Ramachandran, P., Zoph, B., Le, Q.V.: Searching for activation functions. In: NIPS, pp. 4939–4948 (2017)
26. Ruder, S.: An overview of gradient descent optimization algorithms. arXiv preprint arXiv:1609.04747 (2016)
27. Srivastava, R.K., Greff, K., Schmidhuber, J.: Training very deep networks. In: NIPS, pp. 2377–2385 (2015)
28. Tan, M., Le, Q.V.: EfficientNet: rethinking model scaling for convolutional neural networks. In: ICML, pp. 6105–6114 (2019)
29. Vaswani, A., Shazeer, N., Parmar, N., et al.: Attention is all you need. In: NIPS, pp. 5998–6008 (2017)
30. Wang, T., Zhu, Y., Jin, L., Luo, C., Chen, X., Wu, Y., Wang, Q., Cai, M.: Decoupled attention network for text recognition. In: AAAI, pp. 12216–12224 (2020)
31. Wu, Y., Schuster, M., Chen, Z., et al.: Google's neural machine translation system: bridging the gap between human and machine translation. arXiv preprint arXiv:1609.08144 (2016)
32. Yan, R., Peng, L., Bin, G., Wang, S., Cheng, Y.: Residual recurrent neural network with sparse training for offline Arabic handwriting recognition. In: ICDAR, pp. 1031–1037 (2017)
33. Yan, R., Peng, L., Xiao, S., Johnson, M.T., Wang, S.: Dynamic temporal residual network for sequence modeling. Int. J. Doc. Anal. Recognit. **22**(3), 235–246 (2019)
34. Yousefi, M.R., Soheili, M.R., Breuel, T.M., Stricker, D.: A comparison of 1D and 2D LSTM architectures for the recognition of handwritten Arabic. In: DRR, p. 94020H (2015)
35. Yue, B., Fu, J., Liang, J.: Residual recurrent neural networks for learning sequential representations. Information **9**(3), 56 (2018)
36. Yue, X., Kuang, Z., Lin, C., et al.: RobustScanner: dynamically enhancing positional clues for robust text recognition. In: ECCV, pp. 135–151 (2020)
37. Zeiler, M.D.: ADADELTA: an adaptive learning rate method. arXiv preprint arXiv:1212.5701 (2012)
38. Zhang, Y., Chen, G., Yu, D., Yaco, K., Khudanpur, S., Glass, J.: Highway long short-term memory RNNs for distant speech recognition. In: ICASSP, pp. 5755–5759 (2016)

Chapter 5
TH-DL Multilingual Text Recognition System Framework

Abstract By integrating a text detection model with text recognition models, the TH-DL text recognition system framework was designed and implemented for different tasks. Different text recognition models proposed in the book are compared. The experimental results show that PREN2D has achieved the highest recognition accuracy for word-level scene text images, whereas ARN-DTRN has achieved the best performance for sentence-level text images. A multilingual scene text recognition system based on the TH-DL framework was ranked first on the RRC-MLT-2019 leaderboard for the end-to-end text detection and recognition task. An Arabic video subtitle recognition system based on the TH-DL framework ranked first in the ICDAR 2017 and ICPR 2020 competitions on Arabic text detection and recognition in multiresolution video frames.

Keywords Multilingual text recognition · Video subtitle recognition · Text detection · Deep learning · Confidence analysis

5.1 System Framework

The text recognition models proposed in the previous chapters are integrated with a text detection model to construct the TH-DL text detection and recognition system framework, as shown in Fig. 5.1. A confidence analysis module can be introduced after the text recognition model, which is used to control the program flow. By switching to different recognition models, the TH-DL framework can be applied to different tasks.

For text detection, the TH-DL framework adopts a modified Mask-RCNN model [7], which uses ResNeXt-101 [8] as the backbone. The text detection model outputs the coordinates of the four vertices of the text bounding boxes, which determine an arbitrary quadrilateral. The input image is cropped according to the detected text instances, the cropped text images are used as the input of the recognition model. Then, the vertices of the arbitrary quadrilateral are mapped into vertices of a rectangle via bilinear interpolation. The height of the rectangle is equal to the maximum length of the left and right sides of the original quadrilateral, and

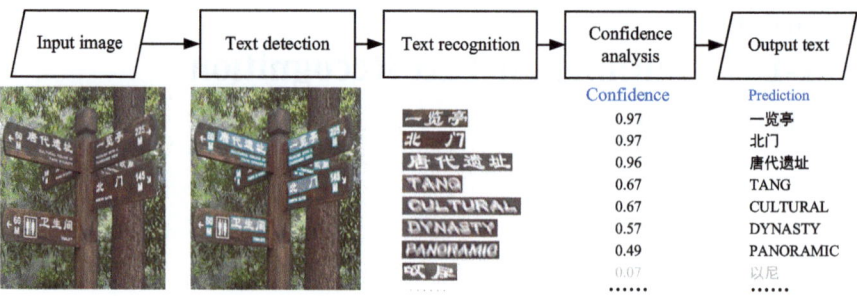

Fig. 5.1 TH-DL text recognition system framework

the width of the rectangle is equal to the maximum length of the top and bottom sides of the original quadrilateral.

For text recognition, different models can be used for various tasks. The recognition model outputs both the predicted text and the confidence of the recognition results. The confidence is calculated as the mean of the Softmax values of all the decoded characters. For images with confidence lower than a preset threshold, the following two steps are performed.

1. **Renormalization and recognition**. If the recognition confidence is lower than the threshold, one possibility is that the normalized orientation of the input image does not match the real reading direction. Therefore, the input image is renormalized in the following three different ways: (1) the image is rotated 90° clockwise and normalized as a horizontal image, (2) the image is rotated 90° counterclockwise and normalized as a horizontal image, and (3) the image is normalized as a vertical image. The three renormalized images are recognized again, and the prediction with the highest recognition confidence is selected as the updated result.
2. **Rejection**. If the highest recognition confidence of renormalized input images is still lower than the threshold, the system will reject the recognition results as the detected region is a false alarm or the text is illegible.

5.2 Model Selection for Different Tasks

To analyze which model is suitable for various tasks, a comprehensive comparison is conducted in this chapter, including the primitive representation learning network (PREN) and its semantic-guided decoding method, the multielement attention network (MEAN), the PREN2D model, and the ARN-DTRN model.

The word accuracy and average time costs of different models on the scene text recognition task are listed in Table 5.1. "PREN (semantic)" refers to a PREN with a semantic-guided decoding module. The average recognition time costs are tested with a single NVIDIA Tesla V100 GPU. As shown in Table 5.1, the PREN has

5.2 Model Selection for Different Tasks

Table 5.1 Word accuracy (%) and average time costs of different models in the scene text recognition task

Model	IIIT5K	SVT	IC03	IC13	IC15	SVTP	CUTE	Time
PREN	91.7	88.9	93.3	94.2	79.7	82.5	84.7	16.3 ms
PREN (semantic)	93.4	92.9	94.9	96.4	82.9	86.5	**88.5**	21.8 ms
ARN-DTRN	95.5	93.2	95.0	95.8	85.1	85.6	87.2	62.4 ms
MEAN	95.8	94.3	95.7	96.6	85.9	87.8	87.1	79.1 ms
PREN2D	**96.5**	**95.1**	**96.3**	**96.8**	**86.0**	**88.1**	**88.5**	47.4 ms

Bold values highlight the optimal performance (e.g., highest accuracy or lowest error rate) for the respective metric across different methods, models, or systems

Table 5.2 Performance (%) and average time costs of different models on the IAM handwritten English word dataset

Model	Val CER	Val WER	Test CER	Test WER	Time
PREN	12.89	25.14	14.57	26.41	16.7 ms
PREN (semantic)	11.43	21.97	13.01	23.24	20.9 ms
ARN-DTRN	8.20	19.68	10.59	22.14	57.1 ms
MEAN	8.01	19.14	10.39	21.74	78.8 ms
PREN2D	**7.45**	**17.73**	**9.44**	**19.80**	45.1 ms

Bold values highlight the optimal performance (e.g., highest accuracy or lowest error rate) for the respective metric across different methods, models, or systems

Table 5.3 Performance (%) and average recognition time of different models on the AcTiV validation set

Model	CRR	LRR	Time
PREN	80.15	57.13	18.6 ms
PREN (semantic)	80.96	58.48	23.8 ms
MEAN	98.43	80.60	446.4 ms
PREN2D	98.05	78.50	237.8 ms
ARN-DTRN	**98.70**	**86.00**	302.2 ms

Bold values highlight the optimal performance (e.g., highest accuracy or lowest error rate) for the respective metric across different methods, models, or systems

the highest efficiency, and the average time to recognize an image is only 16.3 ms. After introducing the semantic-guided decoding method, the model can achieve a significant improvement in accuracy with an inconsiderable additional time cost. Among all the models, PREN2D achieves the highest word accuracy.

The performances of the different models on the IAM handwritten English word dataset are listed in Table 5.2. Similar to that in scene text recognition, the PREN has the fastest recognition speed, and PREN2D also achieves the lowest character error rate and word error rate.

To explore the performance of different models on images with long text lines, the models are compared on the validation set of the AcTiV dataset [9, 10], and the results are listed in Table 5.3. Although PREN has a significantly faster

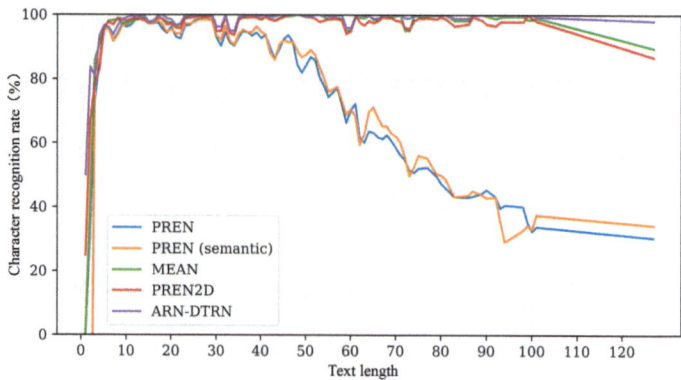

Fig. 5.2 Character recognition rates of different models on images with various text lengths

recognition speed, its accuracy for long texts is limited. One possible reason is that the primitive representations are learned via global feature aggregation, which lacks character positional information. The proposed ARN-DTRN has achieved the highest character recognition rate and text line recognition rate by combining dynamic temporal residual learning with the attention rectification method.

To better analyze the effect of text length on model performance, Fig. 5.2 shows the character recognition rates of different models on images with various text lengths in the AcTiV validation set. The PREN can achieve performance similar to that of attention-based encoder-decoder models when the text length is less than 30 characters. When the text length exceeds 100 characters, ARN-DTRN can maintain stable performance.

On the basis of the TH-DL framework, a multilingual scene text recognition system and an Arabic video subtitle recognition system are implemented. For multilingual scene text recognition, PREN2D is adopted as the text recognition model. For Arabic video subtitle recognition, ARN-DTRN is used for text recognition because video subtitle samples contain long text lines.

5.3 Multilingual Scene Text Recognition System

The International Conference on Document Analysis and Recognition (ICDAR 2019) hosted a robust reading challenge on multilingual scene text detection and recognition (RRC-MLT-2019) [5]. The competition provides a multilingual scene text dataset named MLT19. An additional synthetic dataset containing 277,000 synthetic samples is also provided, which is generated according to the method proposed by Gupta et al. [1]. There are approximately 1.2 million word-level images after cropping the text regions in the original samples according the word localization ground-truth.

Table 5.4 Performance (%) of PREN2D on the multilingual text line samples of the training set of RRT-MLT-2019

Script	Character number	Character recognition rate
Latin	313,930	93.2
Chinese	20,112	77.0
Japanese	31,711	72.8
Korean	23,218	85.2
Hindi	20,613	80.1
Bangla	21,368	75.1
Arabic	23,627	86.1

The evaluation metrics of the end-to-end text detection and recognition task in the RRC-MLT-2019 competition include the F1 score, precision, recall, average precision and character recognition rate. Detailed information can be found in Sect. 1.5.

To support the recognition of multilingual text images, there are two different kinds of approaches. The first approach uses a script classification model that is trained to identify the script of the text. Then, different scene text recognition models for different scripts are used to recognize images in the corresponding scripts. The second approach trains a model for all supported scripts.

In the TH-DL framework, both detection and recognition use only one model to process different scripts. A multilingual scene text recognition system is constructed on the basis of the TH-DL framework, which uses a modified Mask R-CNN [2, 7] for detection and a PREN2D model for recognition. For the training of the PREN2D, because there are many multioriented images, the samples are divided into a horizontal subset and a vertical subset according to the aspect ratios in the training stage. The data of each training iteration are randomly sampled from one of the subsets. In the test stage, the system uses confidence analysis to determine the optimal orientation of an input image.

As the ground-truth of the test set of RRT-MLT-2019 is not available, the performance of PREN2D on the multilingual text line samples in different scripts of the training set is shown in Table 5.4.

The MLT-19 text recognition training set was partitioned into a training subset and a validation subset, with 80% and 20% splits, respectively. The recognition model underwent fine-tuning on the training subset for one epoch, followed by an evaluation on the validation subset. The results, as detailed in Table 5.5, indicate that fine-tuning led to improved character recognition rates in five out of seven languages. Notably, significant enhancements were observed for Hindi and Bangla, likely due to the relatively limited data available for these languages in the original training set.

The proposed system achieves the best performance on the public leaderboard of the end-to-end text detection and recognition task.[1]

---

[1] The link of the leaderboard is https://rrc.cvc.uab.es/?ch=15&com=evaluation&task=4

Table 5.5 Character recognition rate (%) of PREN2D before and after fine-tuning on 80% of the multilingual text line samples of the training set of RRT-MLT-2019 and evaluating on 20% of the multilingual text line samples of the training set of RRT-MLT-2019

Script	Character Number	w/o fine-tuning	w/fine-tuning
Latin	63,085	93.0	**94.4**
Chinese	4221	**75.7**	73.4
Japanese	6655	**73.5**	72.1
Korean	4651	84.6	**84.8**
Hindi	4048	79.8	**84.9**
Bangla	4149	74.3	**82.1**
Arabic	4587	86.1	**86.2**

Bold values highlight the optimal performance (e.g., highest accuracy or lowest error rate) for the respective metric across different methods, models, or systems

Table 5.6 Performance (%) of different methods on the end-to-end text detection and recognition tasks of the RRC-MLT-2019 competition

Method	F1	Precision	Recall	AP	1-CER
mask_rcnn-transformer	51.04	52.51	49.64	25.96	55.71
CRAFTS	51.74	65.68	42.68	34.95	48.27
end2end	52.50	55.34	49.93	40.89	58.47
Tencent-DPPR Team & USTB-PRIR	59.15	71.26	50.55	35.92	58.46
Baidu-VIS	59.72	72.82	50.62	41.32	57.26
CPN (multi-scale)	63.78	67.07	**60.79**	48.87	**67.92**
TH-DL w/o confidence	57.76	61.12	54.75	43.26	63.43
TH-DL	61.76	74.16	52.91	45.58	58.76
TH-DL w/ fine-tuning	**65.15**	**75.47**	57.31	**50.29**	62.67

Bold values highlight the optimal performance (e.g., highest accuracy or lowest error rate) for the respective metric across different methods, models, or systems

The performance of the different systems in the end-to-end text detection and recognition leaderboard of RRC-MLT-2019 is listed in Table 5.6. "w/o confidence" means that the confidence analysis is not used. "w/ fine-tuning" means that the recognition model is fine-tuned on the training set of RRT-MLT-2019 for 10 epochs. The system based on the TH-DL framework has achieved leading performance. Moreover, although the system without confidence analysis has slightly higher recall, both the precision and F1 score are lower. By incorporating the confidence analysis module, the F1 score improved from 57.76% to 61.76%. With recognition model fine-tuning, the F1 score further improved to 65.15%.

Recognition examples of the multilingual scene text recognition system are shown in Figs. 5.3, 5.4, 5.5, 5.6, 5.7 and 5.8. For each sample, the input image is shown in the left column, and the cropped detected regions and the corresponding predictions of the TH-DL system are listed to the right of the image. Wrongly recognized characters are marked in red, and "_" refers to missing characters. Two classic errors are as follows: (1) recognition errors caused by text detection errors,

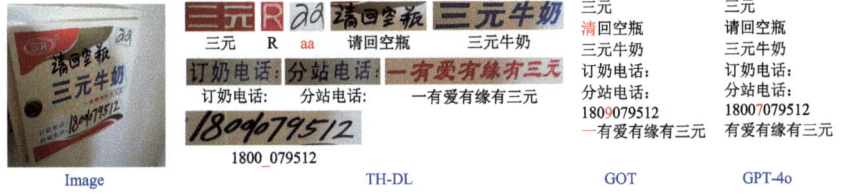

Fig. 5.3 Example of Chinese text recognition in the multilingual scene text recognition system

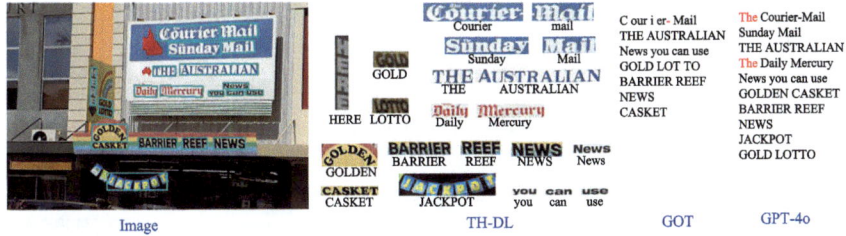

Fig. 5.4 Example of English text recognition in the multilingual scene text recognition system

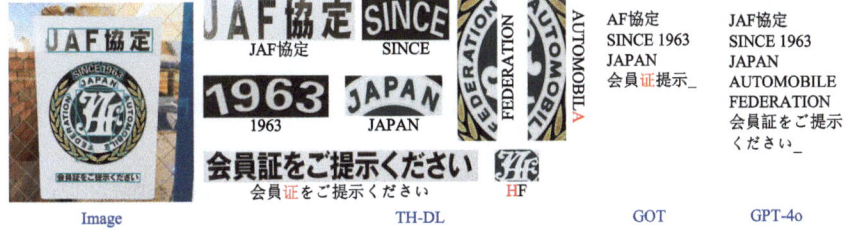

Fig. 5.5 Example of Japanese text recognition in the multilingual scene text recognition system

such as the last character "E" of the word "AUTOMOBILE" in the Japanese sample; (2) similar characters, in which some similar characters only have slight differences; and (3) recognition errors caused by low image quality, such as the character "M" in the word "Mobil", which is recognized as "V" in the Arabic sample. Overall, the TH-DL system can accurately detect and recognize most multilingual scene text images.

Comparisons with multimodal large language model-based systems, including GOT [6] and GPT-4o [3], are also provided in Figs. 5.3, 5.4, 5.5, 5.6, 5.7 and 5.8. The GOT model has 0.58B parameters, including a VitDet [4]-based image encoder, a linear layer, and a Qwen 0.5B-based output decoder. The GOT has strong adaptability to low-quality images and supports Chinese, English, and Japanese. However, the current GOT model does not support other languages. Moreover, its ability to distinguish similar handwritten Chinese characters still needs improvement. In comparison, although GPT-4o benefits from the use of large-scale samples during training and performs well in handling multiple languages, our model can process

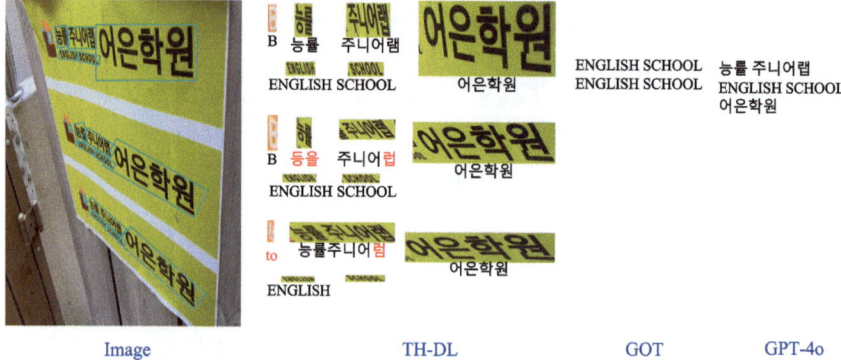

Fig. 5.6 Example of Korean text recognition in the multilingual scene text recognition system

Fig. 5.7 Example of Hindi text recognition in the multilingual scene text recognition system

both horizontal and vertical text in scenes and is better adapted to smaller characters. For example, in the English sample shown in Fig. 5.4, only TH-DL correctly detects and recognizes the vertical English word "HERE". The text detection module in TH-DL has approximately 210 M parameters, whereas the text recognition module (PREN2D) has only 71 M parameters.

5.4 Arabic Video Subtitle Recognition System

Recognition of video subtitles in multilingual scripts is helpful for content-based video retrieval. Arabic video subtitle recognition is a representative task that has attracted researchers' interest. Multiple editions of the competition on Arabic text detection and recognition in multiresolution video frames [9, 10] were organized by the National Engineering School of Sousse, Tunisia, the University of Fribourg, Switzerland, and the University of Applied Sciences and Arts, Western Switzerland. The second and third editions of the competition were hosted by the International Conference on Document Analysis and Recognition (ICDAR 2017) and the International Conference on Pattern Recognition (ICPR 2020), respectively. The publicly available Arabic Text in Videos (AcTiV) dataset contains images collected from

5.4 Arabic Video Subtitle Recognition System

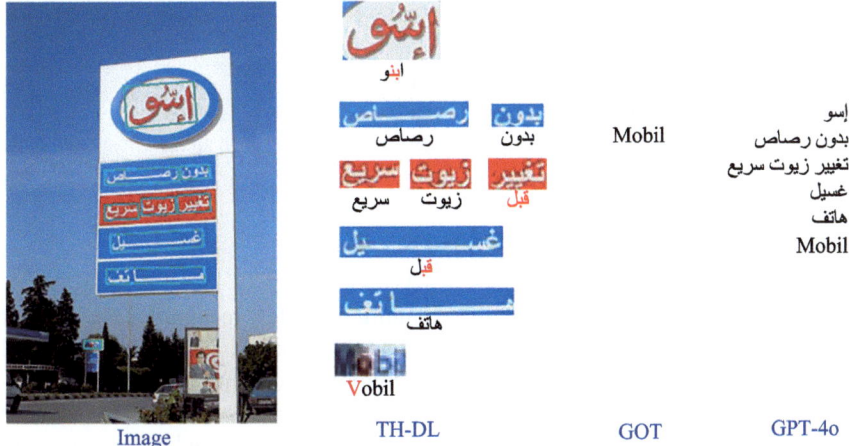

Fig. 5.8 Example of Arabic text recognition in the multilingual scene text recognition system

Table 5.7 Performance comparison (%) of different methods on the AcTiV test set

Method	Metric	HD	SD1	SD2	SD3	Average (SD)	Average (All)
TH-DL	CRR	**99.83**	**99.34**	**99.48**	**99.43**	**99.43**	**99.53**
	LRR	**95.80**	**87.43**	**85.94**	**85.07**	**85.63**	**88.71**
ArabOCR	CRR	99.49	98.31	98.72	99.07	98.75	98.94
	LRR	90.84	72.77	71.48	77.83	74.10	79.03
DCR	CRR	99.67	97.70	98.77	98.77	98.53	98.81
	LRR	89.69	69.63	74.61	58.82	68.26	74.30

Bold values highlight the optimal performance (e.g., highest accuracy or lowest error rate) for the respective metric across different methods, models, or systems

one HD (high definition) channel (AljazeeraHD) and three SD (standard definition) channels (TunisiaNat1, France24 Arabic, Russia Today Arabic).

An Arabic video subtitle recognition system has been implemented using the TH-DL framework. Since video subtitles are usually long horizontal text lines, the CNN-LSTM-CTC scheme is adopted as the recognition model. The TH-DL system won first place in both the ICDAR 2017 and ICPR 2020 competitions on Arabic text detection and recognition in multiresolution video frames [9, 10].

The performance comparisons of the different methods on the AcTiV test set are shown in Table 5.7. The evaluation metrics include the character recognition rate (CRR) and line recognition rate (LRR) [9, 10]. ArabOCR and DCR are other systems that participated in the Arabic text recognition task in competitions at either ICPR 2020 or ICDAR 2017 [9, 10].

The performances of TH-DL CNN-LSTM-CTC and TH-DL ARN-DTRN are further compared on the available AcTiV validation set as shown in Table 5.8. ARN-DTRN has shown better than that of the CNN-LSTM-CTC.

Table 5.8 Performance comparison (%) of TH-DL CNN-LSTM-CTC and TH-DL ARN-DTRN on the AcTiV validation set

Method	CRR	LRR
TH-DL (CNN-LSTM-CTC)	98.59	83.66
TH-DL (ARN-DTRN)	98.70	86.00

5.5 Conclusion and Future Work

This book focuses on deep learning-based multilingual text recognition. In contrast to traditional sequence-to-sequence modeling approaches, such as CTC-based models and encoder-decoder models with attention mechanisms, primitive representation learning is proposed to achieve a balance between recognition performance and inference efficiency. To effectively utilize 2D spatial dependencies and long-term temporal dependencies in text images, the book further explores a multielement attention mechanism and dynamic temporal residual learning with attention rectification for multilingual text recognition.

In the future, it is essential to explore innovative representation learning techniques for multilingual text recognition in challenging scenarios, such as occlusions in open environments. Self-supervised learning and transfer learning continue to be effective approaches that can be further explored to improve the generalization ability of models. With the emergence of large language models, a new research direction has arisen in the development of multimodal vision-language models for various visual text perception tasks. Technologies for artificial intelligence generated content (AIGC), such as diffusion models, can be employed to produce more training samples featuring a comprehensive character set in various styles for multilingual text recognition.

References

1. Gupta, A., Vedaldi, A., Zisserman, A.: Synthetic data for text localisation in natural images. In: CVPR, pp. 2315–2324 (2016)
2. He, K., Gkioxari, G., Dollár, P., Girshick, R.: Mask R-CNN. In: ICCV, pp. 2961–2969 (2017)
3. Hurst, A., Lerer, A., Goucher, A.P., Perelman, A., Ramesh, A., Clark, A., Ostrow, A., Welihinda, A., Hayes, A., Radford, A., et al.: GPT-4o system card. arXiv preprint arXiv:2410.21276 (2024)
4. Li, Y., Mao, H., Girshick, R., He, K.: Exploring plain vision transformer backbones for object detection. In: ECCV, pp. 280–296 (2022)
5. Nayef, N., Patel, Y., Busta, M., Chowdhury, P.N., Karatzas, D., Khlif, W., Matas, J., Pal, U., Burie, J.C., Liu, C.l., et al.: ICDAR2019 robust reading challenge on multi-lingual scene text detection and recognition – RRC-MLT-2019. In: ICDAR, pp. 1582–1587 (2019)
6. Wei, H., Liu, C., Chen, J., Wang, J., Kong, L., Xu, Y., Ge, Z., Zhao, L., Sun, J., Peng, Y., et al.: General OCR theory: towards OCR-2.0 via a unified end-to-end model. arXiv preprint arXiv:2409.01704 (2024)

References

7. Xiao, S., Peng, L., Yan, R., An, K., Yao, G., Min, J.: Sequential deformation for accurate scene text detection. In: ECCV, pp. 108–124 (2020)
8. Xie, S., Girshick, R., Dollár, P., Tu, Z., He, K.: Aggregated residual transformations for deep neural networks. In: CVPR, pp. 1492–1500 (2017)
9. Zayene, O., Ingold, R., BenAmara, N.E., Hennebert, J.: ICDAR2017 competition on Arabic text detection and recognition in multi-resolution video frames. In: ICDAR, pp. 1460–1465 (2020)
10. Zayene, O., Ingold, R., BenAmara, N.E., Hennebert, J.: ICPR2020 competition on text detection and recognition in Arabic news video frames. In: ICPR, pp. 343–353 (2020)